HOME

AM
MO

HOMEMADE AMMO

How to Make It, How to Reload It, How to Cache It

DUNCAN LONG

PALADIN PRESS • BOULDER, COLORADO

Also by Duncan Long:

AK47: The Complete Kalashnikov Family of Assault Rifles
The AR-15/M16: A Practical Guide
AR-15/M16 Super Systems
Assault Pistols, Rifles, and Submachine Guns
Combat Revolvers: The Best (and Worst) Modern Wheelguns
Combat Rifles of the 21st Century: Futuristic Firearms for Tomorrow's Battlefields
The Complete Ar-15/M16 Sourcebook: What Every Shooter Needs to Know
Hand Cannons: The World's Most Powerful Handguns
The Mini-14: The Plinker, Hunter, Assault, and Everything Else Rifle
Mini-14 Super Systems
Modern Ballistic Armor: Clothing, Bomb Blankets, Shields, Vehicle Protection
. . . Everything You Need to Know
Modern Combat Ammunition
Modern Combat Blades
Modern Sniper Rifles
The Poor Man's Fort Knox: Home Security with Inexpensive Safes
The Ruger .22 Automatic Pistol: Standard/Mark I/Mark II Series
Streetsweepers: The Complete Book of Combat Shotguns
Sturm, Ruger 10/22 Rifle and .44 Magnum Carbine
Super Shotguns: How to Make Your Shotgun into a Do-Everything Weapon
The Terrifying Three: Uzi, Ingram, and Intratec Weapons Families

Homemade Ammo:
How to Make It, How to Reload It, How to Cache It
by Duncan Long

ISBN 0-87364-816-1
Printed in the United States of America

Published by Paladin Press, a division of
Paladin Enterprises, Inc., P.O. Box 1307,
Boulder, Colorado 80306, USA.
(303) 443-7250

Direct inquiries and/or orders to the above address.

Contents

Warning

Technical data presented here—particularly technical data on ammunition and the use, adjustment, and alteration of cartridges for various firearms—inevitably reflect the author's individual beliefs and experience with particular firearms, equipment, accessories, and components under specific circumstances that the reader cannot duplicate exactly. Therefore, neither the author nor the publisher assumes any responsibility for the use or misuse of information contained in this book.

Procedures in this book and the resulting end product can be extremely dangerous and should be approached with the greatest of caution by only those capable of handling the task and who have taken measures to protect themselves from accidents. Procedures in this book may also be illegal according to local, state, or federal laws. Therefore, readers should contact the proper authorities before attempting any reloading or ammunition fabrication.

For these reasons, the information in this book is for *information purposes only*.

Acknowledgments

Thanks must go to the many reloaders, manufacturers, importers, dealers, friends, and others who fed me information for this book and offered advice and equipment. Given today's climate, all must remain anonymous.

Thanks must go, too, to the fine people at Paladin Press for going out on a limb and publishing this book. My usual very special thanks must go to Maggie, Kristen, and Nicholas for their help.

My prayer is that none of these people, or any of my readers, will ever need to put into use what is detailed on the following pages.

INTRODUCTION

Government Stimulation

I started this book at the beginning of 1994, partly as a result of what has taken place during the previous year.

It was a year in which the gun grabbers not only called for outright bans on many firearms and magazines with capacities of more than 10 rounds, but also managed to pass the so-called Brady Bill. This was followed in October by the "Crime Control" Bill that outlawed 19 specific types of firearms—even though government statistics clearly indicate that these particular weapons are seldom chosen by criminals.

In 1994, the U.S. Surgeon General outlined the need for "safe weapons . . . safer guns and safer ammunition"(whatever in the world these might be). Congressmen aligned with the Brady Bunch also suggested we ban hollow-point ammunition because it is more apt to kill criminals. Though, in truth, full-metal-jacket bullets are much more dangerous because they dictate filling a criminal with lead in order to stop him—and increase the penetration of bullets, making it more likely innocent bystanders will be injured in the process.

Sen. Patrick Moynihan (D-NY), during one of his more sober moments, went so far as to propose a 1,000 percent tax on ammunition often used for self-defense. Winchester answered the challenge by restricting one of its more effective 9mm rounds to police use and preventing citizens from buying it.

After easing the top Bureau of Alcohol, Tobacco, and Firearms officials into an early (and paid) retirement following their actions in Waco, U.S. Treasury Secretary Lloyd Bentsen suggested we should crack down on criminals by raising the Federal Firearms Licensing fee for gun dealers to $600. Included in the recommendations from the "greatest minds in our country" was a proposal from several congressmen to repeal the Second Amendment.

It doesn't take a modern-day Daniel to read the handwriting on the wall. Those with the luxury of having bodyguards (most often at taxpayers' expense) are out to disarm those of us who walk the mean streets brought about by politicians' ill-conceived criminal laws.

Much of the shooting public has received the message. One of my friends who operates a gun store reported that over the first three weeks of December 1993, he sold more guns and ammunition than during all the previous five years—showing that President Clinton has stimulated at least one segment of the economy for the time being. By the end of December, the price of assault rifles (now banned despite the fact that they are rarely used in crime) had rocketed to four to five times their normal retail price. And most stores had their shelves stripped of the more popular ammunition.

There are millions of guns in private hands in America, and it's doubtful that any sort of government roundup of them will be entirely successful. Even if it were, alcohol prohibition during the 1920s showed that resourceful basement tinkerers will soon be peddling black-market wares. Tomorrow's bootlegger will be selling steel and ammunition instead of gin.

Firearms technology, despite today's plastics and stainless steel, hasn't changed much over the last 100 years. Almost any handyman with access to a drill press, a hacksaw, and some files can produce a submachine gun in a few weeks. (Pakistanis, working with 19th-century tools, have produced an infinite variety of firearms this way, and many of the guns drew Soviet blood in Afghanistan.)

Guns can last almost forever. If you have a firearm that you can hide from government gun grabbers, a little grease and oil will keep it like new for decades. You'll probably even be able to hand it down to your children or grandchildren if you manage to keep it hidden from government agents.

Ammunition is another matter.

Although the parts for a submachine gun aren't for the most part hard to fabricate, and can often be found in their basic forms at a local plumbing and hardware stores, most people don't know the first thing about the chemistry behind making smokeless powder. And even if you do, where would you go to buy the components for your gunpowder? A savvy pharmacist or chemical supply clerk will know right away what you're up to and might report you to the authorities.

Even if you had a keg of powder, you may not know the first thing about reloading cartridges with it. And what if you don't have bullets available? What about primers? How can you know whether brass can be reused or not? Teaching you how to reload ammunition—even if the number of components on hand is zero—is what this book is all about.

Of course, you might simply stock up while cartridges are still legal to buy and own. But even if you are wise (and wealthy) enough to do this, you might use up your stock during one of the firefights that will most likely result as crime and anarchy follow in the wake of the disarmament of honest people. Or maybe someone will break into your house and steal your ammunition. Or perhaps the government will discover your cartridges and take them from you, throwing you into jail for a couple of years to teach you a lesson in political correctness.

Obviously, knowing how to create ammunition from scratch could save your bacon if you're interested in defending yourself.

At the same time, this isn't a book about how to reload ammunition by using standard components (there is already a number of good books on the subject, and most manufacturers supply everything you need to know with their reloading kits). So I'm not going to waste any space going into the how-tos, which powders are best, etc. But there are a few things to keep in mind, which are covered below.

THE BASICS

First, the key to reloading is having some sort of reloading die. This is necessary to compress the brass cartridge, which expands slightly during firing. A die enables you to resize the brass back down to its original specifications, so that its neck so it will hold a bullet in place, and decreasing the cartridge's girth so that it will chamber easily. The press itself can also be used to swage or resize bullets and other objects, thanks to its compound leverage.

You can get by without a press, but it is tough. In the past, shooters have discovered they can use the chamber of a firearm to resize the empty brass. To do this, the chamber has to be cleaned meticulously. Then the cartridges are lubricated (to keep them from getting stuck) and placed into the chamber. The bolt is slammed down onto the cartridge, driving it into the chamber to resize it. Of course, this puts a lot of wear and tear on the extractor of the bolt, and the cartridge is only resized slightly. And because the neck of the cartridge isn't narrowed in this process, it's necessary to hold the bullet in place with shellac, tar, or some other improvised adhesive. These cartridges have to be placed into a chamber by hand because the bullet is likely to be forced back into the cartridge during recoil if it is chambered from a magazine. But in the wilds of Africa and elsewhere, shooters have been able to make do with this method of resizing brass they discover in the bush.

If you're forced to reload ammunition without a press, several other tricks make the job easier. The best substitute for a press is a common bench vise. By filing down a nail so THAT it will fit through the primer hole inside the brass, it is possible to place an empty round in the press with the outside of the primer just above the top edge of the jaws. When the nail is positioned inside the brass with the head of the nail resting on the opposite jaw of the vise, simply tightening it will drive out the spent primer.

Another method of doing this is to file down the nail and place it inside the brass with the cartridge sitting on a nut and the primer over the center opening of the nut. Lightly pounding on the nail with a hammer will drive the primer out the base of the cartridge. Simply drill a small hole in a good solid bench and let the spent primers drop through into a container.

After the primer has been reloaded (as outlined below), it can be reseated with a vise. To do this, the brass is placed in the vise with the primer positioned over its hole, resting against the jaw of the vise with the mouth of the brass against the other jaw. Slowly tightening the vise will ease the primer into its hole and seat it. By working carefully to avoid crushing the brass, reloaders can reseat primers with a vise.

A hammer and a flat, wooden surface can also be used to reseat a primer once you cut a small piece of dowel to fit through the mouth of the cartridge. To do this, the primer is placed on the surface and the brass positioned over it with the dowel in the mouth of the brass. Tapping very lightly on the top of the dowel will gradually drive the brass downward, seating the primer. (This work will scar the surface of the wood you're working on, leaving tell-tale signs of what has been done—so be careful.)

Of course, using shop tools to prime/deprime cartridges is an operation of last resort, but if you have to reload without a press, it can be done. That said, you'd be wise to purchase a reloading kit now while they're still available rather than having to exploit a firearm or shop tools for reloading brass.

There's a huge variety of reloading outfits sold. They vary from small kits of dies that are used with a plastic mallet and can be carried in a pocket to full-fledged factory-style "progressive" presses that cough out a loaded round with each pull of the handle.

The pocket-sized reloading kits are neat, but they are very slow. Because powder, bullets, and primers aren't easily portable and reloading generally requires a workbench, the apparent benefit of a small kit is also nearly nonexistent, especially when the slow reloading pace that such kits demand is taken into consideration.

So you're better off getting at least a hand press that will take the labor out of the process and speed things up. If you're serious about reloading, you'll need to get a press with several "stations" on a turret that permits mounting dies and rotating them into position as they're needed. This speeds up the job and also does away with the need to adjust the height of the dies each time you reload a particular caliber.

A cartridge will get jammed in a die if the brass isn't lubricated before it's resized. Don't forget this. The only exception is carbide dies, which are slick enough to prevent this. Carbide dies are also a bit more rust resistant, though you should still exercise care in their storage.

The best resizing lubricant is obviously that from the manufacturer. Lee Precision, Inc. offers an excellent water-soluble lubricant that is easy to clean off reloading dies and cartridges after it has been used and has the added plus of not deactivating powder. Old-timers used about every imaginable lubricant under the sun with varying results. Automotive chassis grease seemed to have been best, though it created dents in cases if too much was applied. A similar grease is a good bet if you're improvising for your reloading work.

If you go with an oil-based product that may damage powder (and most oils and greases will), then use graphite or molybdenum disulfide to lubricate the inside of the mouth case before running it through the resizing die. This will keep down excessive wear on the expanding plug, as well as keeping damage to the brass to a minimum. Polymerized oil such as STP Engine Treatment or Motor Honey is a very effective lubricant for resizing or swaging (dampen a strip of thick felt and roll the cases or bullet blanks on it), but it must be thoroughly removed with a petroleum solvent (naphtha or gasoline) before the rounds are reloaded.

When you start shopping for dies, you'll also discover that some resize "full-length" empty brass, whereas others just resize the neck of the cartridge. The latter takes considerably less work, but also may jam in a dirty or tight chamber or, if fired in a gun other than the one used previously, may be oversized for this different chamber. Either can be disastrous if you're using the firearm for self-defense. So your best bet is to use a full-length resizing die to guarantee that your brass will be super-reliable.

(About the only exception to this is a gun that has been fired excessively; such a firearm may have excessive headspace, causing brass to separate upon firing when it has been full-length resized. In such a case, you can back the die out of the press slightly to increase the headspace while still keeping the chambering reliable even if the chamber of the firearm becomes dirty.)

There are a number of manufacturers offering quality reloading equipment. My favorites are the assemblies offered by Lee Precision, Inc. because the company offers a lot of quality at a very reasonable cost. The company also supplies its reloading kits with all the information you need to use various powders with the cartridge dies; smaller kits even come with a small

A turret press (like this one from Lee Precision) makes reloading quick and easy.

A die set permits reloading cartridges for a specific chambering. A different set of dies will be needed for almost every cartridge you plan to reload.

measuring cup to fill cartridges quickly without having to weigh powders.

Lee's low-end products include the Hand Press and Reloader Press, which hold only one die at a time. But they're small and cost under $25, allowing you to get into the reloading process without a huge investment. A two-die kit for reloading rifle ammunition will run around an additional $55, and a three-die pistol die kit costs about $60. If you select one of these presses, then plan on sending as many shells as you need through each die once it is placed on the press and adjusted. Then do the same with the next die, and your final step will be to seat the bullet after you're finished with the dies. This gets around the slower pace dictated by a single-position press and minimizes the time lost in exchanging and adjusting the dies.

For just a bit more, you can purchase a turret press that has a rotating die holder. This type of press permits

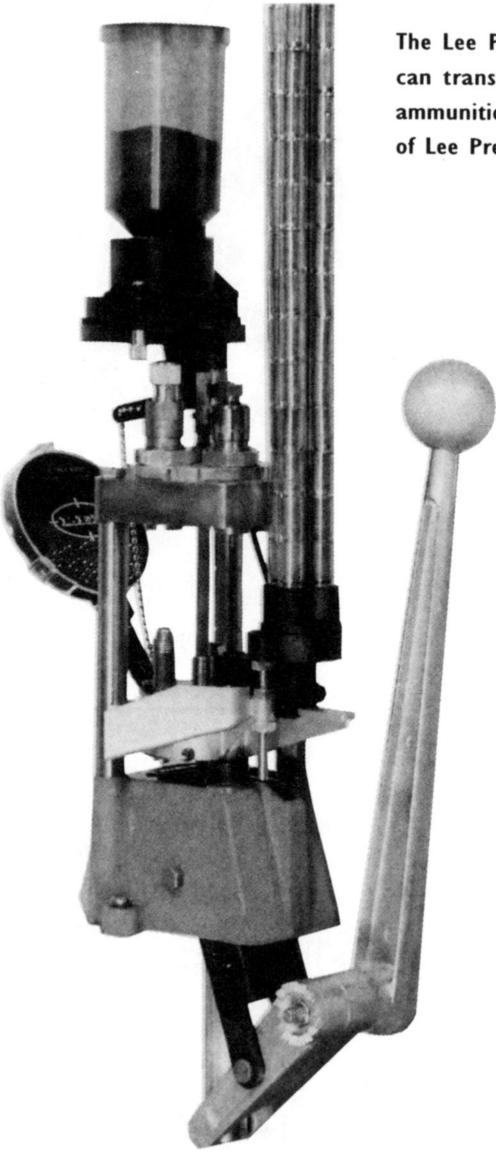

The Lee Pro-1000 progressive press can transform you into a one-man ammunition factory. (Photo courtesy of Lee Precision, Inc.)

reloading just a few shells at a time without having to readjust the dies each time you use the press. Lee's turret press allows a quick exchange of the turret and its dies so you can rapidly switch from reloading one type of ammunition to another without readjusting the dies (once you've done this task for the first time and locked the dies in place).

The cost of a Lee turret press is around $68 (or $82 with an auto-index attachment that rotates the dies into the next position with each pump of the handle). Lee offers extra turret die holders for around $11 each.

For really chugging out ammunition the Lee Load-Master is ideal; this assembly can load both rifle and pistol rounds (with the conversion from one to the other pretty quick, thanks to snap-out die turrets that permit leaving the dies in place so they don't need to be readjusted). The Load-Master automatically inserts a case and advances it, positions the primer, drops in a powder charge, and seats a bullet on the cartridge. All you do is load the storage bins on the machine, pump the handle, and watch to ensure that everything is going as it should. A loaded cartridge is spit out with each crank of the handle.

The suggested retail price for the Load-Master is $189, plus the cost of the dies and shell holder (also available from Lee), or approximately $330 for a complete kit of dies and press that practically allows you to start loading ammunition as soon as you take the gear out of the box.

If you're only interested in reloading pistol ammunition, then the Lee Precision's Pro-1000 is an excellent choice. It has most of the automated features of the Load-Master, but reloads only pistol cartridges and carries a slightly lower price tag. As does the Load-Master, this progressive press turns out a reloaded cartridge with each pull of the lever once you get it started. The cost for this reloading kit is around $189, and the kit includes the dies for the 9mm Luger cartridge, so there's nothing else to buy other than powder, primers, and bullets in order to start reloading. The Lee kits also come with clear instructions to get you started.

When a cartridge is chambered, the bullet often hits the feed ramp or edge of the chamber with a lot of force. If the bullet isn't firmly in place, it can jam back into the cartridge, thus creating excessive pressures when it is fired. Therefore, it's essential to get good crimps on cartridges; bullets with cannelures are ideal, though these bullets can be tough to find and hard to make.

Fortunately, Lee offers a crimping tool that both improves accuracy and keeps the bullet in position during chambering. The crimper works like a die and is easily mounted in a press. It costs around $18, making it a good investment.

Another trick that might improve the reliability of

ammunition is to chamfer the outside rim of the mouth of the brass to help prevent the edge of the cartridge neck from getting hung up on the front edge of a magazine during chambering.

This is a good idea on "bottle-neck" cases that headspace on the shoulder; however, most rimless, straight-case autoloading pistol rounds headspace on that square front lip of the brass case. If you taper (chamfer) its leading edge on the *outside*, the case will tend to feed too deeply into the chamber/bore: if it goes too deeply in the chamber/bore, the firing pin may not be able to reach it, and it won't fire. If the firing pin is long enough to hit it, odds are fair that it will go off—and tend to stick in the chamber to cause extraction problems. Most people slightly chamfer the *inside* of the case, to help the bullet go in straight, but only trim

Shotgun ammunition comes in a variety of lengths and gauges. Shown here are three of the more common. At the left is a 3-inch Magnum 12-gauge shell. The center shell is a standard length 12-gauge Magnum. The shell on the right is a 20 gauge.

to length and do not chamfer the outside of straight-rimless pistol rounds.

One final piece of equipment that will turn your operation into a real ammunition factory is a lead smelter coupled with some bullet molds. This permits transforming scrap lead from tire weights or other sources (but not lead batteries since the acid in these makes them dangerous to melt) into bullets. Although in a pinch you can melt lead in an iron pot suspended over a campfire, an electric lead pot makes the job a lot easier. Lee offers a near 110-volt Production Pot IV for around $55, as well as molds in most of the popular calibers for around $24 for a two-cavity mold (producing two bullets at a time). The handle for the mold runs another $15 or so (and one handle can be used with any number of molds since it is more like a pair of pliers than an actual handle). And if you want to go into full-scale production, Lee Precision even offers six-cavity molds for $50, making it possible to really go to town.

If you'll be reloading shotgun ammunition, the

A huge number of popular cartridges are available for handguns. If you are going to obtain the cartridges you need during a government crackdown, you'll most likely have to reload them yourself.

required equipment is even less expensive. The Lee Load-All II comes with everything you need to load one gauge of shot shells and costs just about $50 for the complete kit of dies, powder scoop, primer seater, and so forth. The kits are available for 12-, 16-, or 20-gauge ammunition. The Lee Load-Fast 12-Gauge Press puts out shells at twice the rate of the Load-All II (giving the Load-Fast a top rate of around 200 shells per hour); it costs about $155, and the primer feed costs an additional $20 or so.

All of Lee's kits come with complete instructions on how to reload, determine how much of which powders to use, etc. But for a more detailed look at reloading, see my *Combat Ammo of the 21st Century* (available from Paladin Press), as well as Robert S.L. Anderson's *Reloading for Shotgunners,* Dean A. Grennell's *ABC's of Reloading,* and Edward A. Matunas' *Metallic Cartridge Reloading.* Ken Warner's *Handloader's Digest,* currently edited by Bob Bell, is also a good place to find what new equipment is available in the reloading marketplace. (All except *Combat Ammo* are available from DBI books or in gun stores offering reloading supplies.)

Once you become familiar with reloading, you may wish to invest in a brass polisher, neck trimmer, and other specialized equipment. But don't buy this equipment until you get your feet wet. You may discover that there is a lot of stuff you can get by without.

1 Primer Directive

As you probably already know, the primer is the part of a cartridge that is struck by the firing pin, generating a small blast of hot gases that ignite the main powder charge inside the round. The primer is pretty simple in operation, but its smallness makes it hard for the do-it-yourselfer to manufacture or recharge. Therefore, if you're still able to purchase these on the open market, you should consider stocking up *now* so that you'll have them if a ban of or excessive tax on reloading components ever gets through the legislature.

Of course, the size of primers works in your favor if you purchase them. They're easily concealed and transported since you can hide a hundred of them in a tiny container. Primers are arguably a smuggler's dream come true.

Like powder, primers are sensitive to heat, moisture, light, and lubricants. Therefore, you need to keep both your powder and primers dry. Also store primers in a dark, cool place and be sure not to expose them to any type of oil or other lubricant.

There are basically two styles of primers: the most common one in the United States is the Boxer primer, while the Europeans, ironically enough, use the Berdan primer originally invented by an American (go figure). The Berdan primer is tough to remove because it fires through twin holes rather than a larger central port (as the Boxer primer does).

Some American reloaders manage to remove the primer by filling the case with water and ramming a rod down the mouth of the cartridge; the hydraulic pressure pops the primer out. This is messy and time consuming. It should be noted that the Berdan primer is also used on cases that manufacturers know won't survive more than one shooting. Therefore, if you find an aluminum or steel cartridge with Berdan primers, don't try to reload it because the case will most likely rupture on the second firing.

Currently, you'll encounter primers in a variety of sizes: small pistol, large pistol, small rifle, and large rifle; shotgun ammunition has primers similar to those in

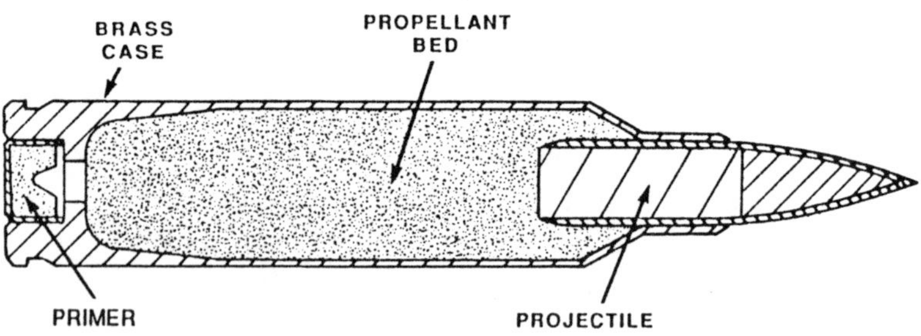

Cross section of rifle cartridge. Pistol cartridges are nearly identical to this cartridge except for the straight walls and wider bullets found on most such cartridges.

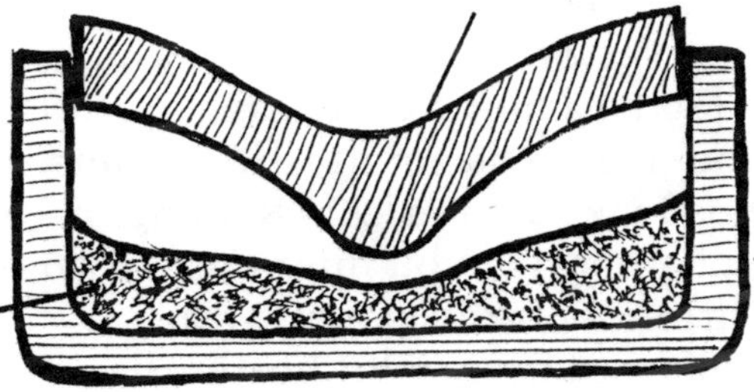

Cross section of Boxer primer.

rifles, but the "battery" style is self-contained, with the anvil striking the front of the primer rather than the base of the cartridge. (European brass cartridge shotgun shells often use rifle primers—but you're not likely to encounter any of these.)

Unfortunately, large rifle and pistol primers are nearly the same size as small rifle and pistol primers. Although this allows you to reload rifle ammunition with pistol primers in a pinch, the results can be disastrous. Some semiauto rifles with a floating firing pin (like the AR-15) may fire when the bolt closes on the cartridge—*not* a safe situation. All types of rifles also have heavier firing pins that may pierce pistol primers, sending a cloud of hot gas back into the action—and maybe into the shooter's face. Likewise, pistols may lack enough force to fire ammunition that has rifle primers in it. So plan on using pistol primers in handguns and rifle primers for rifle ammunition unless you have no other choice than to experiment with unorthodox combinations of components.

If you have to choose between buying small rifle (or small pistol) primers or large ones, purchase the smaller ones because they're the hardest to reload. Shotgun primers are the largest and therefore the easiest of all primers to reload. Too, since shotguns are generally reserved for close encounters, only a few loaded shells are probably ever going to be used with a shotgun.

On the other hand, an assault rifle might conceivably go through several hundred shells in a short time if the shooter is facing a motorcycle gang or rioters. For this reason, stocking up on small rifle primers that can be used to reload the 5.56mm/.223 Remington cartridge (or whatever chambering you might use in your personal defensive firearm) makes a lot of sense.

Magnum primers are a bit hotter than the standard primer and are used with powders that sometimes don't burn reliably with standard primers. Because any powder you use might be old or of poor quality, you should consider purchasing magnum primers, *provided* they can also be used for the cartridges you regularly reload. A quick check of the company's reloading literature or a call to the company will help you out here. (Just be sure to get the information while it is still available. If the government starts closing things down, there will be a run on components, accompanied by a slamming of the doors to knowledge.)

All right—suppose you've found a batch of empty cartridges with primers in place. How do you reload the primers?

It's important not to rush things since primers are basically small explosive charges; mistakes can create serious injuries. The process is also very exacting and time consuming.

During the reloading process, you'll knock out the empty primer when you resize the cartridge (with most reloading kits). Most reloaders simply toss the empty primers. But if you'll be reloading them, you need to save each one. Since it is hard to reload primers, it also doesn't hurt to have a lot of spares.

When it comes to reloading primers, commercial ones are easier to recycle than military ones. Military primers are seated in cartridges with a crimp, making it hard to remove them without damage. By contrast, commercial primers pop out easily—and it is even possible to unload a live primer without damaging it: a trick you might want to keep in mind if you trade for a cartridge for which you have no firearm.

When you start working with live primers or commence creating chemicals for reloading old ones, it is important to remember that *primer chemicals are explosive.* It's essential that you exercise extreme caution when reloading primers. Never keep large amounts of the chemicals used for reloading primers in the area where the primer is being seated.

Also, always wear eye protection and protective clothing when working with primers (or during any other reloading task). This rule is doubly important if you elect to reload your own primers. And if you have any doubts about your abilities, it's better not to tackle the job than to injure yourself. Reloading primers is dangerous business, at best, and not for those who are inexpert with explosives.

The hard part of reloading primers, of course, is to create the explosives that will arm them. Once you've done this, as is outlined later, the basic operation for reloading a primer is not that hard. Just do the following:

1. Clean the primers of carbon deposit and dirt. If you have it, spray carburetor cleaner works well.
2. Remove the anvil (the three-pronged, flower-shaped metal inside the primer) from the primer cup. This is most easily done with a sharp-pointed tool.
3. Clean the primer cup and then use a small tool to flatten out the dent placed in the bottom of it by the firing pin of your gun. (A hammer with a flat-ground nail works well for this.)
4. Pour the primer chemical into the cup, compact it in the base of the cup, and allow it to dry.
5. Place the anvil in the primer pocket of the brass

that's to be reloaded (the brass should be cleaned and polished before the anvil is placed in it).

6. Place a tiny paper disk or very thin piece of foil over the primer chemical. The size and thickness of the material can be determined by a little experimentation. The disk may not be necessary if the powder grains are large enough that they will not flow through the primer hole; if the powder is fine, the disk keeps the powder from clogging the primer and blocking the anvil from reaching the primer's explosive. You might also place a thin disk of foil or paper over the primer hole for this same purpose; a little glue will be needed to keep the cover in place, and it must be very thin to permit the "flash" from the primer to ignite the powder in the cartridge.
7. Reseat the primer cup into the brass.

Once this is completed, you have a rifle or pistol brass ready for the powder charge and bullet. Shotgun primers can also be reloaded. The main difference with shotgun primers is that they also contain a "built-in" primer pocket on the primer rather than in the hull itself. This means that you have to dismantle the shot shell primer after it has been removed from the hull. (A .223 Remington depriming tool is about the right size for this task.)

The anvil in the shotgun primer is shaped like a Y rather than like the three-sided flower of the pistol and rifle primers. Other than the extra steps involved in removing the primer from its removable pocket and then replacing it, the procedure is much the same as for the brass cartridge. It's also *essential* that the paper disk be placed on the shotgun primer to keep powder from clogging the primer.

Reloading centerfire primers is tough; reloading rimfires is a real pain in the posterior. Generally, it isn't worthwhile to reload .22 rimfire brass, but if it means the difference between being unarmed or having a working firearm (or if the prices of ammunition have catapult-

Reloading .22 rimfire ammunition is really tedious. However, the small size of many hide-out .22s may justify the extra work needed to reload these cartridges. Shown here is a small Sterling automatic pistol.

ed bootlegging ammunition into a lucrative new profession), you might want to take extra pains to do so.

In rimfire cartridges, the primer goes inside the rim. When the two sides of the rim are smashed together by the gun's firing pin, the primer ignites and sets off the powder in the rest of the cartridge.

Now, you could go to a lot of trouble and create a tool to iron out the original firing pin indent in a .22 cartridge. But this isn't really necessary because the chances of the firing pin striking the exact spot twice is small and, even if it does, if you've placed new primer chemicals in the rim of the shell, it will likely fire anyway. So don't worry about the little dent made by the first firing of the shell. In the heyday of shooting galleries—when .22 shorts cost less than a penny apiece—reloading machines for .22s were common. Designs varied, but they were usually smaller versions of factory loaders that held the case upright for loading a small amount of fluid priming compound, then spun it rapidly enough for the compound to be forced to the hollow rim and held there until it dried to the desired firmness. In this manner, .22 cases could be reloaded many times.

Here is the procedure for repriming an empty rimfire:

1. Clean the old primer residue from the inside of the rim of the brass with a small tool.
2. Dampen the primer chemical with alcohol, acetone, or methyl ethyl ketone (MEK)—you'll have to experiment to see which works best. You want it to be *almost* liquid, but not runny, so that you can place it in the rim and then have it stay there.
3. When you have the right consistency, place the damp primer material in the cartridge and use a tool to push it into the rim area of the brass. One other way of getting it into the rim is to place the cartridge, base down, into a drill (power drills are best if you have electricity available to run them) and spin it at 1,000 RPM or so for 30 seconds.
4. This done, place the brass in a warm place to allow the primer to dry.

I've heard of people using toy caps (as used in cap guns) for black-powder primer caps. These might also be placed in large rifle or large pistol primer cups to reactivate them. But given the corrosion these caps produce, cleaning them thoroughly, as outlined in Chapter 5, is essential. However, they might be a possibility if things get bad. Just cut around the powder charge of the paper and then lacquer it into place inside a bit of aluminum foil, courtesy of Reynolds.

More durable black-powder firearm primers can be made from heavy aluminum foil and the primer chemicals outlined below. The gauge of aluminum that used to be found in frozen dinner packages works well for these primers; this aluminum foil is a bit scarce today, but a little searching will usually turn it up. Once you find a few pieces, squirrel them away for later use because black-powder weapons will undoubtedly become popular following a government ban on smokeless-powder guns because of the lack of serial numbers on most black-powder pistols, rifles, and shotguns sold in the United States.

Here's the procedure for making a black-powder primer:

1. Use the nipple of the black-powder firearm as a form to shape a square of aluminum over the primer.
2. Trim off the excess aluminum so it is sized like a standard primer.
3. Coat the inside of the primer cup that you've formed with a primer chemical (described below).

Because most of the primer chemicals listed below are hygroscopic, it's a good idea to keep your black-powder primers in a sealed container until you're ready to fire them. You might be able to place a light coat of lacquer or other waterproofing chemical over the inside of the primer, but this can make ignition a bit iffy. So experiment carefully before adopting such a practice.

MAKING PRIMER CHEMICALS

In the past, the easiest way to secure primer chemicals was to collect the white tips of strike-anywhere matches and then dissolve their chemical so it could be painted into empty primers.

Unfortunately, safety-conscious match manufacturers, with an eye toward lawsuits (and perhaps a bit of government pressure) adopted less explosive chemicals for the tips of matches. This lessens not only their effectiveness as matches, but also as primer material.

If you diskover matches that will work, great. It's easier than making the chemical. You can determine whether the match tips will work by smacking them smartly with a spoon or small tool. If the strike ignites the match, the chemical is ideal for a primer. If no amount of tapping will ignite it, then it isn't going to work as a primer (though, as we'll see in a bit, the match head can be employed as a component for several types of powder).

The following steps can be used to create primers if you find suitable matches. Once again, *primer chemicals are explosive.* You have to be careful, or you will ignite the chemicals. Wear safety glasses and keep amounts of chemicals you collect in any one sitting very small to minimize both the danger and your losses if you set the works off. Here're the steps:

1. Carefully break the white tips off the strike-anywhere matches, dampen them with alcohol, acetone, or MEK, and then grind them on a hard, flat surface. Usually a piece of glass works well for this.
2. Once the amount you need has been powdered, mix it with a little alcohol to create a paste. If alcohol doesn't dissolve the mix, try acetone or MEK.
3. Place the paste in the primer cup.
4. After the mix has dried completely, the primers are ready to be placed into cartridges.

If you find matches (or other chemicals) that can be used for primers or powder, you need to exercise a little caution in purchasing more of the materials. If you go through the checkout counter at your local store and buy 8,000 matches, someone might start to get a bit suspicious, especially if government agents have alerted stores to report such occurrences. So use a little sense and go about things in a cautious manner.

Prepare a logical explanation for any primer chemicals you purchase. If possible, it is also wise to pay with cash and purchase chemicals where people don't know you. Given the universal use of TV security cameras, it might not be too paranoid to even consider disguising your features somewhat, such as adding or switching glasses, changing your hairstyle, or donning a hat or cap.

Even after the goodies are at your workplace, you still must be careful. The fumes created in making powder or primers can be toxic and corrosive as well as highly explosive. Extreme care must be taken to avoid injuring yourself or burning down your house. Usually, the best bet is to work outdoors or to have a good exhaust system (and a kitchen exhaust fan will *not* suffice).

There are many ways to modify primers to create different burning rates. Unfortunately, most of these take a lot of experimenting to get right, waste a lot of materials, and attract unneeded attention. Because of this, it's generally best to make all-purpose mixes. And once you find a combination of components and chemicals that works, don't try to modify things. If the system works, don't change it.

Not all the chemicals suitable for primers are covered here. Silver permanganate, antimony sulfide, barium nitrate, red phosphorus, calcium silicide, or tetranitroaniline are sometimes employed. Barium nitrate is corrosive and often hard to obtain—and would be even harder to obtain in a regulatory environment where all nitrates and oxidizers were controlled. But various barium salts are widely used in industry. To make barium nitrate, get some barite (barium sulfate mineral), which is fairly cheap and is used as a weighing mud in drilling oil wells; or if you want a finer grade (it doesn't matter) get some from your favorite internist who uses it as the opaque medium in gastrointestinal X-rays (the famous barium milkshakes you hear about). Mix it with ground coal and roast to high red heat in an iron furnace, which will give you barium sulfide. Wash out the barium sulfide with hot water. Add an excess of nitric acid to this and you will have barium nitrate solution. Filter, evaporate, and you have barium nitrate.

Red phosphorus is noncorrosive but burns on exposure to air, meaning that it must be mixed under a nonoxidizing liquid such as kerosene. Additionally, primers must be zinc-plated in order to protect them when they are filled with red phosphorus.

For these reasons, potassium chlorate is often the best bet for do-it-yourself primers because it's easy to make, even though it dictates extra cleaning of any firearm that it's fired in.

Although manufacturers sometimes mix primer chemicals, it is essential that you don't experiment with them because you can create dangerous, unstable results. Red phosphorus and potassium chlorate, for example, detonate upon contact with each other. Leave mixing operations to the big manufacturers.

You can add relatively inert materials to primer chemicals to increase the friction and/or burning temperature when they're ignited. Aluminum powder or glass powder is best for this. Aluminum increases the burning heat, whereas glass increases the friction created when the primer is struck by the firing pin. If you decide to experiment with these materials, they should be ground into fine powder for best results.

Potassium Chlorate

Potassium chlorate is not a primer material by itself but can be used with other chemicals to create a primer. However, potassium chlorate is very sensitive and and requires careful handling. Be careful when mixing it with other materials, especially with certain sulfur/sulfide and phosphorus compounds because it might contribute to an explosion. Do take all precautions and wear safety appropriate safety gear.

Potassium chlorate is also corrosive, causing rust in a firearm within a matter of hours. For this reason, potassium chlorate isn't the best choice for gas-operated firearms. If you do use potassium chlorate, then be sure to follow the cleaning instructions outlined in Chapter 5 of this manual.

To make potassium chlorate, you'll need potassium hydroxide (lye), hypochlorite solution (usually in household bleach—check the label), and sodium bisulfate (in most toilet bowl cleaners). One catch: there are two forms of lye; the one you want is potassium hydroxide, not sodium hydroxide. Generally, a stroll down a supermarket aisle will give you the ingredients you need—be sure to read the fine print in the ingredient section of the label of all products.

The process of changing potassium hydroxide into potassium chlorate involves substituting its chlorine group for the "hydroxide" in the potassium compound. The equipment you need is a Pyrex bottle with a one-hole stopper, rubber or glass tubing to connect the gas coming from the bottle to a bath, a large glass baking pan, a fine cloth or filter paper, and pHydrion paper (available from most drugstores and some gardening stores—try the latter first and tell the clerk that you're testing your soil for acidity). If you can't find pHydrion paper, litmus paper or even vinegar can be used in a pinch.

Make only small amounts of the final primer material at a time because this chemical is *highly* explosive. Here's the procedure:

1. Make a bath of lye solution in a glass pan (do not use aluminum; the lye will react with it). Use one part—by weight—of lye to two parts of water. *Caution*: Add the lye *very slowly* because when it dissolves it generates enough heat to cause the water to boil and/or splash. Wear eye protection and have a bottle of vinegar handy to neutralize any spills.
2. Mix the sodium bisulfate (several spoonfuls) into the bleach and then put the mixture into the bottle.

3. Cork the bottle and vent the gas through the tubing so that it bubbles through the lye solution. The gas produced is poisonous; *be sure you vent it fully.*
4. When the lye solution is transformed into a potassium chlorate solution, it will change from a base into an acid. You can determine when the process is finished by testing the solution with the pHydrion or litmus paper. If neither of these is available, you can test the solution by taking a few drops in an eyedropper and putting it into vinegar. Base chemicals, such as vinegar, fizz when placed in acids. When the solution quits reacting with the vinegar, the solution has become acidic.
5. Pour the acidic solution into a shallow pan and place a small flame under it. *Do not allow the solution to boil vigorously.*
6. Watch for crystals that will form on the bottom of the pan. When these start to form, turn off the heat and allow the solution to cool. This will cause the potassium chlorate to precipitate from the solution, forming more of the crystals.
7. When the crystals have precipitated, pour the liquid and crystals through a cloth, paper towel, or other filter. Save the crystals trapped in the filter and diskard the liquid.
8. Dissolve the crystals in water, 1 part crystal to 1 part water.
9. Pour the mix into a shallow pan and place a small flame under it. *Do not allow the solution to boil vigorously.*
10. Continue heating until crystals again form on the bottom of the pan. When this starts to happen, turn off the heat and allow the solution to cool so that the potassium chlorate again precipitates from the solution.
11. When the crystals have precipitated out, pour the

liquid and crystals through a cloth, paper towel, or other filter. Keep the crystals trapped in the filter and diskard the liquid.

13. Grind the damp potassium chlorate crystals to the size needed while they are wet.
14. Once the powder has dried, mix it with a little alcohol to create a paste. If alcohol doesn't dissolve the mix, try acetone or MEK.
15. Place the paste into the primer cup.
16. After the mix has dried fully, it's ready to be placed in the cartridge.

Mercury Fulminate

Making mercury fulminate is a simple procedure but calls for mercury and nitric acid, both of which are usually hard to obtain (although, as we'll see later in the powder section, it's possible to make the nitric acid).

The major downside to mercury fulminate is that it is hard on brass. After the primer ignites, it deposits mercury on the inner surface of the brass, with the pressure of the burning powder driving it into the brass. Once the mercury amalgamates with the brass, the metal becomes brittle over time.

The embrittlement of the annealed (softened) brass can lead to split cases or head separation in instances where the chamber and cartridge fit isn't right on the money. However, old-timers used mercuric primer routinely since it was the only one available. In a bare-hands situation, mercuric primer is the safest, surest homemade primer to make and use. And if you use nickeled cases, the mercury will have no effect. You can get nickeled cases in any handgun caliber. And if you're still worried, for a few bucks you can get a bottle of electroless nickel compound and nickel-plate the inside of all your brass cases.

On the other hand, mercury fulminate doesn't cause

much damage to a firearm, unlike potassium chlorate, which is such an effective oxidizer that it promotes rust in firearms within hours of being fired. So if you're reloading for a semiauto or selective-fire gun that spews its brass everywhere—making it difficult to find for reloading anyway—then mercury fulminate is a better choice for primers because it won't cause excessive rust problems, even if it does ruin the brass cartridges.

Just remember that if you settle on mercury fulminate for primers, your cartridges may become one-shot affairs. This means you'll need a good source of brass if you're going to continue reloading. It also means you need to mark your cartridges with a hash check on the cartridge head so that you don't get them mixed up with good cartridges. This will also prevent having someone gather up your old brass and try to barter it back to you. Because of the potential danger to reloaders, you should smash empties that you collect. You certainly shouldn't leave them out in the open, if it's at all possible, where they're easily picked up (and it may not be possible for you to do this if you're using the cartridges for combat).

This preamble out of the way, here's the procedure for creating mercury fulminate:

1. Mix 10 parts of mercury to 74 parts (by measured volume) of nitric acid.
2. When the mercury has dissolved, heat the solution to 130°F.
3. Slowly pour the solution into 100 parts (by measure) of alcohol. *Be careful: splashing can create dangerous burns or fires.* An effervescent reaction will occur, and white fumes will be given off. *The fumes should be vented because they are dangerous.*
4. When the effervescent reaction stops, filter the liquid through several paper towels or through filter paper.
5. Rinse the particles in the filter with cold water.

6. While they're still damp, *very carefully* grind the particles into a fine powder.
7. Place the powder in a pan and dry it by placing the pan in a second filled with boiling water.
8. Once the powder has dried, mix it with a little alcohol to create a paste. (You can also use either acetone or MEK.)
9. Place the paste into the primer cup.
10. After the mix has dried fully, the primer is ready to be placed in the cartridge.

• • • • •

After the primer is seated in the brass, it's a good idea to seal the primer in place with a shellac or lacquer, sealing around the outside rim of the primer to ensure that moisture doesn't get into the primer. Don't forget to do this; most of the do-it-yourself primer chemicals are hygroscopic.

Additionally, after firing a gun loaded with any of these primers, it's essential to clean it immediately because many of these primer chemicals are corrosive or hygroscopic. Follow the procedures outlined in Chapter 5 for cleaning a firearm.

2

Powder Power

Modern smokeless powder is difficult and dangerous to make. Therefore you should try, if it's at all possible, to purchase or otherwise obtain powder rather than try to manufacture it yourself.

By studying the reloading charts and manuals offered by powder manufacturers (and you should get a collection of these because they can be worth their weight in gold), you'll discover that some powders can be used for a variety of cartridges and a few can be employed for loading both shotgun and pistol ammunition. The trick is to find the cartridges you're most likely to be reloading, then cross-reference the loading charts to locate the powders the various cartridges have in common. Once you identify them, try to purchase as much of the powder as you are likely to need and then store it as outlined below.

It's also possible to obtain powder from cartridges similar to those you will be reloading, pull the bullets from the cartridges, and use the powder (and maybe the primer as well) in your rounds. However, this can be dangerous if it is not approached with caution and some common sense. But if you check reloading manuals, you can compare the amount of powder normally used in the cartridges you've found to those you're wanting to reload for. A little "guessometry" and a powder scale or even a scoop can enable you to transfer the powder in the proper proportion for your cartridge and then you can reload it using your found components. Of course, you must be very conservative when substituting powder; it's better to have weak rounds than cartridges that blow the barrel off a firearm.

Don't be tempted to use powder of unknown origin or that which has been loaded into a dissimilar cartridge, mortar shell, or what have you. For example, you could experience disaster if you tried to load rifle cartridges with powder scavenged from pistol rounds.

Regardless of the source, proper storage of powder is essential if it is to last. Because the solvents left in many powders are necessary to preserve their power, it's important to keep them in sealed metal containers. Don't place powder in glass containers; light quickly damages powder.

Heat is another enemy of powder because it dries the moisture in the powder quickly and causes it to deteriorate. Once this process occurs, the powder ages quickly, undergoing chemical changes that will eventually ruin it. Storing powder in an attic or a hot toolshed will make it unsuitable for use in a short time. Cold storage of powders can be ideal; however, a constant temperature is better than one that dips to very low temperatures and then climbs to more moderate ones.

If proper storage is used, you can expect powder to last for 20 years or, if kept in a refrigerated area at a constant temperature, up to 50 years.

Rust-colored dust on the surface of some powders indicates that they are deteriorating. Other bad signs are corrosion on metal parts of containers and the replacement of the ether/alcohol smell of the powder by an ammonia odor. When any of these signs are found, the powder should be discarded. If the old powder is in cartridges, the brass and bullets can be salvaged, provided the acid by-products of corrosion haven't actually attacked the brass and copper. Chances are that the primers will be unusable if the powder is bad, but you might want to test the primers to see if they can be salvaged.

If the primers are still good, they can usually be popped out with a depriming tool or the resizing/depriming punch on a reloading press (the only exception to this rule is military brass with crimped primers). When removing live primers, care should be taken to operate well away from powder, other primers, or flammable materials. And, of course, it is important to wear eye protection as in other reloading operations.

Don't be tempted to use powder to load cartridges once it has deteriorated. In addition to not firing consistently, it may create a "squib" firing that lodges a bullet in the barrel. With a semiauto or selective-fire gun, this contributes to inadvertently sending another shot down the barrel and blowing up the weapon. You're better off not risking this.

If you manufacture your own powder by following the steps below, remember that some types of powder that deteriorate may also generate enough heat to create spontaneous combustion. For this reason, you should store the powder in an area away from your house, check it from time to time, and dispose of any that is no longer good by scattering it over the ground in an open area where it can decompose (and act as a good fertilizer in the process). It's possible to burn old powder, but the fumes can be toxic and the process is almost guaranteed to attract unwanted attention.

There are a number of powders that can be manufactured for cartridges. All are dangerous to varying degrees; the greatest room for disaster lies, however, in manufacturing modern powders. One mistake with these and you can create a serious explosion, fire, or acid spill. Making powder is extremely dangerous, even more so than making primers.

You'll also face a lot of unknowns if you're making your own powder. Although small grain sizes are ideal for pistol ammunition and larger grains for larger rifle ammunition, sizing is relative and varies greatly from one cartridge to another. Too, shotgun ammunition tends to vary greatly, with slower burning powders being used for large bores and faster powders for the .410 and smaller bores. Because some powders are inherently faster burning than others, regardless of the size, figuring the best size for any given gun is tricky and requires a lot of cautious experimentation. All of this necessitates an overly cautious and conservative approach if you are to survive the process.

Some of the more common propellants you might use in a firearm are black powder, ammonpulver, cordite, guncotton, and potassium nitrate mixtures. Often, combinations of these are found in commercial powders, but it is generally wise not to experiment with mixtures too much since you may create a dangerous powder in the process.

Storage of do-it-yourself powders is even more of a problem than with commercial powders because most described below are hygroscopic to some extent may decompose more quickly than commercial powder, and some—especially black powder—are more akin to explosives than propellants. Care should be taken to keep your powder in a cool place and in a tightly sealed container designed to rupture if exposed to a fire (to prevent the container from becoming a bomb!). You can't be too careful in making, storing, and using gunpowder or smokeless powders.

POTASSIUM NITRATE

The simplest powders to make are those created from potassium nitrate (also known as saltpeter). The most common is created by mixing potassium nitrate with sugar; the best ratio is 7 parts (by measure) of potassium nitrate to 6 parts of sugar. This potassium nitrate powder works well with rifles .50 caliber and under. It doesn't work well in shorter barrels like those found on most pistols or submachine guns, but does work in long-barreled carbines that employ pistol cartridges.

Droppings from bats, birds, or the like, as well as manure from barnyards or even human latrines, are potential sources of potassium nitrate. However, for decomposition to take place, the temperature must be around 100°F, and large amounts of time and dung are needed for the process to occur. Thus, natural sources of potassium nitrate are relatively rare and are usually restricted to caves or islands in warm climates.

A more roundabout route can hurry the process by using calcium to chemically bind the nitrate in excrement and then exchange potassium for the calcium in a second step. To achieve this end, lime is first placed on a stable floor, outhouse pit, manure pile, or old burial grounds (for the stout of heart), and the area can be used for some time.

Months or even years later, the pit can then be "mined," and the earth-excrement-lime mixed in water so that calcium nitrate is dissolved from the mix. When the water is removed from the material that settles to the bottom of the mixture, the calcium nitrate is dissolved in it; boiling the water leaves behind the calcium nitrate mixed with salts.

Wood ash (which contains potassium carbonate) is added to water, and the calcium nitrate crystals dissolved into the solution. The potassium and calcium exchange places, creating calcium carbonate (slaked lime) and potassium nitrate. The water solution is filtered to remove the calcium carbonate, which is more or less solid; the potassium nitrate is dissolved in the water so it will pass through a fine filter.

You can create a filter from paper, laboratory filter paper, coffee filters, or similar products. Since these fine filters tend to clog, you should also create a "prefilter" of coarser material. Prefilters of choice in the 1800s were made of straw, which still works well. Cloth or other material may give better results, and your best results come if you filter the water a number of times to get rid of as many impurities as is practical.

To retrieve the potassium nitrate, boil the water until it's nearly all gone. Before the water is totally evaporated, remove the heat source and allow the water to evaporate by leaving it exposed to the air (otherwise you may damage the crystals). The crystals left behind will be primarily potassium nitrate.

Here's a step-by-step procedure that begins with the creation of a filter to contain the earth and nitrate:

1. Get a bucket and punch holes in its bottom.
2. Place a cloth over the bottom of the bucket and then stack straw over it and another layer of cloth. Next, place a layer of straw over the cloth and add a third layer of cloth, topping it with about an inch of wood ash.
3. Place a large container under the bucket.
4. Fill the bucket with your source of nitrate and earth.
5. Pour boiling water through the soil and let it drip into the container below the bucket. Pour the water very slowly and use only 1 part water for every 2 parts earth mix.
6. Allow the water to stand for a number of hours and then drain the water off and discard any sediment that may have gotten through the filter.
7. Boil the water for two hours. While the water is boiling, small crystals of salts will form on the bottom of the container; remove these and discard them.
8. When half the water has boiled from the solution, let it cool for 30 minutes.
9. Add alcohol to the solution (1 part alcohol to 1 part solution) and pput the mixture through a filter made of a paper towel. Crystals of potassium nitrate will be left in the filter.
10. To purify the potassium nitrate, redissolve it in clean water and boil the water for two hours. While the water is boiling, small crystals of salts will form on the bottom of the container; remove these and discard.
11. Again add alcohol to the solution (1 part alcohol to 1

part solution) and put the mixture through a filter made of paper towel. Crystals of potassium nitrate will be left in the filter.

12. Allow the crystals to air dry and then seal them in a waterproof container until needed.

Once an ample supply of potassium nitrate has been created, it's necessary to mix it with sugar to create the actual powder.

Here's how to do that:

1. Mix (by volume) 56 parts of potassium nitrate with 48 parts sugar in 84 parts of water. (Up to 3 parts ferric oxide—rust—is sometimes added to increase the burning rate of the powder.)
2. Boil the mix over a small flame while stirring it. Both the potassium nitrate and sugar will dissolve in the water.
3. Boil the water down to one-fourth its original volume. This should create a thick "fudge."
4. Pour the mixture onto a flat surface and allow it to dry in the sun, if possible. Scoring deep furrows in it will speed the drying process.
5. When the mix is moist to the touch but no longer sticky, granulate it by pushing it—a little at a time—through a mesh or screen.
6. Allow the particles created to dry in the sun.

The size of the granules will be determined by the size of the screen. Ideally, you should use a fine window screen for this process. If you have a mechanical bent, you can devise a press similar to that used on pasta machines and to extrude the moist mix through small holes to create long threads that can be broken after the mix has dried. This would give you very precise control of the burning rate of your powder, permitting you to increase the diameter of the strands to slow down the burning rate or decrease the diameter to speed it up.

Your "powder" is ready to load once it has dried. If you store it, put it in an airtight container. Don't store the powder in glass because it will deteriorate more quickly if you do. After you load cartridges with this powder, carefully seal the cartridges with lacquer (as discussed elsewhere) since the mixture is highly hygroscopic.

POTASSIUM CHLORATE

Potassium chlorate is a substitute for black powder as well as a primer material. Like black powder, potassium chlorate is highly corrosive and dangerously explosive. That means that if you use it in a firearm, it's essential that you clean it within a few hours of firing, or you're likely to see some serious rust form on your firearm. And you *must* use a minimal amount of this powder to avoid blowing up a firearm.

When potassium chlorate is used for making powder, it must be mixed with sugar to down its burning rate. The mixture can then be substituted for gunpowder with slightly less of the potassium chlorate being loaded than for the same charge of black powder. In a bind, this material can also be substituted for "smokeless" powder in cartridges, but may not create enough energy to cycle the bolt of semiauto weapons. Again, to avoid excessive chamber pressures, take great pains to not use too much powder.

Potassium chlorate is found in most match heads. Safety matches are almost pure potassium chlorate; if you are using strike-anywhere matches, be sure to remove the tip of the match. Failure to remove the striking tip could create excessive chamber pressures.

You'll need a candy thermometer or similar device to keep track of the temperature of this mix. Here's the procedure to create the gun powder substitute:

1. Mix 1 part table sugar with just enough water to dissolve it into a slush and heat it slowly to 250°F until the sugar melts. Take care not to let the sugar turn brown (carmelize) from excessive heat.
2. Remove the sugar from the fire and stir it as the temperature drops.
3. When the sugar has dropped to 150°F, add 1 part potassium chlorate to the "fudge" a little at a time as you mix it into the sugar
4. Pour the mixture onto a flat surface and allow it to dry in the sun, if possible.
5. When the mix has dried to the consistency of fudge, granulate it by pushing it through a mesh or screen. (If the mixture is hard, too much heat was applied to the sugar; if the mixture is gooey, too little heat was used. In either case, discard the mix and start over, paying careful attention to the temperature.)

The size of the granules determines the burning rate of this mixture. Small granules should be used for pistols or shotguns, whiereas larger sizes are suitable for muzzle-loading rifles. As noted above, you might also fashion a mechanism to create small strands of this powder for more precise control of burning rates.

Your "powder" is now ready; place it in a sealed container or load it into cartridges. Be sure to seal the cartridges carefully because the mixture is highly hygroscopic. And don't forget to clean your firearm after using this powder.

BLACK POWDER

Black powder is hard to make and dangerous as well because it ignites very easily. That means that unlike modern smokeless powder that just smolders if lit, a scoop of black powder left on a workbench or on the floor becomes a potential firebomb that can be set off by a static electricity spark or even by friction. Always store black powder in small containers of a pound or less capacity and that are kept separate from each other. Because water neutralizes black powder, you can wash down an area where black powder has been spilled to make it safe to clean up.

At first glance, sulfur would appear to be unnecessary for the manufacture of black powder. But, in fact, sulfur is needed to lower the ignition temperature of black powder so that it will ignite more readily. Sulfur also improves the consistency of black powder (which is a mix of chemicals and not an actual chemical compound).

If you have trouble securing sulfur, then you may wish to try making brown powder, discussed below, since it requires only a very small amount of this chemical. Or you might want to simply mix potassium nitrate with sugar to create the powder described above (which is basically black powder with the sulfur left out).

The exact proportions of chemicals in black powder vary from one country to the next. In the United States, it tends to be 74 percent potassium nitrate, 15.6 percent charcoal, and 10.4 percent sulfur; the "traditional" mix is 75/15/10. It's also possible to create another version of black powder by substituting and reducing the amount of sodium nitrate for potassium nitrate. However, this type of black powder is deliquescent and especially hard to store in areas with high humidity.

Manufacturers used to coat the grains of black powder with graphite. This was achieved by tumbling it in the final stages of manufacturer. You can do this, too, but should take great care that this is done in a "sparkless" container with no potential for causing an accidental ignition of the powder or containing pressure. (This technique can also be used with other of the powders listed in this section.)

Black powder does have some important pluses. First of all, in black-powder arms it can be used by volume, making it easy to recharge a weapon without having to do a lot of careful measuring. This is an important plus that is somewhat offset by the slow-to-reload designs of most black-powder guns. Another big plus with black powder is that it actually becomes better with age. Provided that it's kept free of moisture, it will not deteriorate with time or when stored in a warm environment. That means that if you have made some black powder (or purchased some beforehand), it will be better than new when you need to use it if you've kept it in an airtight container.

After firing a weapon charged with black powder, it's necessary to clean it and the brass cartridges (if any) thoroughly in order to keep them from corroding. Failure to do this will quickly ruin your firearm.

To make black powder:

1. Obtain potassium nitrate (see above) and carefully grind it into a very, very fine powder.
2. Get some charcoal and grind it to very fine powder. (Note: not all charcoals are created equal. Evergreen charcoal is all but unusable. Dogwood, willow, and alder—in that order—are the charcoals of choice, and they must be thoroughly charred, all aromatics driven off, and free of ash.)
3. Grind sulfur into powder.
4. Mix the three powdered chemicals in the following amounts (by weight): 75 parts potassium nitrate, 15 parts charcoal, and 10 parts sulfur.
5. Place water into the powder and mix it together thoroughly. As long as the mixture is wet, it will not explode.
6. Allow the mixture to dry into a cake after compacting it. Air drying will not work if the humidity is very high; in humid areas, place the mix over a *very* low heat, without exposing it to a flame. Keep in mind that excessive heat will create an explosion.
7. Once the cake is thoroughly dry, carefully break it into chunks with a plastic or wooden tool. Work only with small amounts at a time, with no containers of

the material nearby. Extreme care must be exercised because once it is dry, black powder is highly flammable.

8. Grind the chunks up, one at a time. Although the majority of the powder and the chunks are protected from ignition, take extra precautions to protect yourself, since you could ignite it accidentally during grinding. A screen or sieve is ideal for the grinding process. Fine powder is suitable for pistols; coarser grains should be used for rifles. Experiment with light loads to determine what degree of coarseness or fineness of granules works best.

BROWN POWDER

Brown powder made its appearance just as smokeless powders were being introduced. Technically, brown powder is a form of black powder that was developed for long-barreled, rifled artillery pieces. It was usually called "cocoa powder," due to its color and the fact that it was made in flat wafers that looked like chocolate candy. It's stretching things to say it was close to smokeless powder in effectiveness—it came and went pretty fast. Its salient feature was that it was slow to ignite, which gave a more gradual push to a cannon projectile, of benefit in a long-barreled artillery piece. It gave higher velocities only in very long-barreled weapons. It can be used as a substitute for smokeless powder, but it is far from perfect in small arms.

It does raise a cloud of smoke when it is fired, just as black powder does. Consequently, it came in second in the contest with smokeless powders and was soon discontinued, with manufacturers continuing to make the widely accepted black powder for use in the antique arms that shooters continued to use.

Brown powder has many of the pluses of black powder, including the ability to be stored for long periods without deteriorating. Its additional bonus is that it offers a higher velocity and can be used as a substitute for smokeless powder.

The principal change in the composition of brown powder is that it employs underburned (brown rather than black) charcoal and a greatly reduced amount of sulfur. The proportions of brown powder aren't as firmly established as those of black powder, but it would appear that a 78.5 percent potassium nitrate, 18.5 percent brown charcoal, and 3 percent sulfur mix would be a good place to start. Traditionally, incompletely charred rye straw was used for the charcoal because this type of charcoal acted as a colliding agent, which allowed the shaped pellets of cannon powder to be more densely formed, thus making it a slow-starting powder and better for long-barreled cannon.

Because brown powder is a mix like black powder rather than a chemical combination, the steps in combining the ingredients and creating grains from a cake of the material as given above for black powder should be used in the manufacture of brown powder.

AMMONPULVER

Ammonpulver is made from ammonium nitrate fertilizer. This chemical is most commonly found in home garden and agricultural fertilizers and is readily available in most areas. Wood alcohol is also needed to process the fertilizer, and charcoal is mixed with the chemical to create the proper burning rate.

Ammonpulver had a spate of success as a military powder for use in cannons about the time of World War II and was a close second to smokeless powder in power. It had an Achilles' heel, however: it had to be pressed into very large hollow grains to control the speed of ignition, but at about 30°+C the crystal structure changed, and the large grains crumbled to a powder. In its powdered state, it ignited so fast that the pressure would build up and the guns (artillery) blow up. This makes it less than ideal except in the direst extreme.

Because many fertilizers have calcium coatings on the particles or are mixed with agricultural chemicals, you must exercise a little care in the selection of the fertilizer to be used in manufacturing powder. The best way of determining what you have is to check the ratio of nitrogen, phosphorous, and potash in the mixture; this ratio will be clearly marked on the bag. The proper ratio for making ammonpulver is 34-0-0. If the bag reads 21-0-0, 21-44-8, 46-0-0, or any other ratio, it won't work. Be sure the fertilizer gets all its nitrogen from ammonium nitrate and has no urea added. Urea also dissolves in alcohol, so you'd end up with a mix of ammonium nitrate/urea. Also check to be sure that calcium hasn't been added to coat the mixture. (If you can find only a fertilizer with calcium, crush it to expose the inside of the prills, and it will work just fine.)

You should manufacture your powder immediately after opening fertilizer: it quickly pulls moisture from the air, rendering it useless for making powder after a while. For this reason, you should be careful to purchase new fertilizer that is still sealed in its bag.

When you're ready to start, do the following:

1. Place the fertilizer in a pan and pour methanol alcohol (wood alcohol) over it until the fertilizer is completely covered. The mix should be warm, but do not apply heat to it (i.e., don't mix it outside in the dead of winter or build a fire under the pan—just work in a comfortable environment).
2. Stir the mix until most of the fertilizer is dissolved into the alcohol. (If there is any calcium coating on the fertilizer, this may take some time.)
3. Pour off the alcohol and save it; discard the sludge left behind.
4. Cool the alcohol by placing its container into an ice bath or, better yet, into a second container of dry ice. This will cause crystals of pure ammonium nitrate to form in the container.

5. Run the alcohol and crystals through a filter. (The alcohol can be recycled to retrieve more ammonium nitrate from another load of fertilizer.)
6. Place the crystals in the sunlight and allow them to dry.
7. If the crystals are not small enough, use extreme caution in grinding them since they can explode.
8. Mix the crystals at a rate of 8 to 9 parts crystals with 1 to 2 parts charcoal.

This mixture is highly hygroscopic and should be kept in sealed containers, and the ammunition should be sealed as well. Always clean firearms after shooting this ammunition to prevent corrosion of the barrel and other parts.

NITROCELLULOSE (GUNCOTTON)

The procedure for making nitrocellulose is extremely dangerous and should not be attempted by anyone other than a skilled chemist working in a well-equipped lab. In fact, when this material was created in the mid-1800s, experimentation was suspended for some time because the results were always the same: factories and labs exploded, and all the workers were killed. You can't make any mistakes with this material, or you'll be maimed or dead.

Making smokeless gunpowder is a fairly complicated business. Whether you end up with a propellant or an explosive depends on very subtle nuances in the manufacturing protocol, materials, density, and treatment of the final product and myriad additives designed to stabilize or modify burning characteristics. Such things as extruding techniques, solvents, final solvent content (a function of drying protocols), temperatures, final density, additives, coatings, and make a drastic difference in the final burning characteristics. There have been thousands of smokeless powders made in the past hundred years—and they're all different . . . yet chemically very similar. The differences are as much a result of manufacturing technique as they are of chemical variations. Relatively subtle differences can result in a blank or very fast pistol powder, or a slow rifle powder. And a fast pistol powder will likely blow up a rifle. *Do not attempt to make smokeless powder unless you are qualified to do so.*

To carry out these operations, both sulfuric and nitric acids are needed. Both are highly corrosive, and only their containment in glass prevents dangerous reactions. They are hard to neutralize. Gloves, safety goggles, lab apron, and hooded ventilation systems are *essential* for this work.

In addition to sulfuric and nitric acids, you'll also need a very pure source of cellulose for making nitrocellulose. The best source is probably cotton.

Wood pulp is a fine source of cellulose, and it makes good nitrocellulose. The Brits and Europeans tended to favor cotton; Americans preferred wood cellulose. But for the sake of simplicity, use surgical cotton (making sure it really *is* cotton, because many of the cotton balls used for removing cosmetics are now made from synthetics).

Because of the danger presented by these materials, make only small amounts of this powder at a time and store it well away from the working area.

Since you may not be able to find nitric acid, we'll take a look at how to create it with sulfuric acid. If you have both chemicals, then skip the next section. Diluted sulfuric acid is commonly found in vehicle batteries. To concentrate it, boil it until white fumes appear. These fumes are dangerous, so adequate ventilation is essential. When the fumes start, lower the heat as the acid is ready.

To make nitric acid, follow these steps:

1. Prepare dry potassium nitrate using the method given above.
2. Place 2 parts of the potassium nitrate crystals into a lab bottle with a stopper with one hole. Place a piece of glass tubing with rubber tubing attached to it in the

hole. Don't fill the bottle more than one-fourth full of the crystals.

3. Add 1 part sulfuric acid to the bottle and mix with a glass rod until the crystals form a paste.
4. Run the rubber tubing through an ice-water "bath" and into a second open bottle. Place the stopper in the bottle with the potassium nitrate-sulfuric acid paste in it.
5. Place the paste bottle over a gentle heat. This will create brown nitrogen dioxide and water vapor fumes that will be condensed back into a liquid when they go through the bath. The nitric acid will then drip into the second bottle. Take extreme care not to breathe the caustic fumes that do not condense in the bath and be sure that you have adequate ventilation.

Extreme caution must be exercised in making nitrocellulose. Do not attempt this if you lack proper equipment or knowledge. The procedure is as follows:

1. Immerse the cotton in a mixture of 1 part sulfuric acid and 1 part nitric acid. Allow the cotton to become saturated and then remove it and store it in a sealed container for 24 hours in a cool area.

   Prepare the acid mixture by *slowly* pouring the sulfuric acid into the nitric, while constantly stirring. This will create an exotherm, and the acid mixture must be allowed to cool to room temperature before proceeding.

   Specific gravity of the nitric should be equal to or less than 1.4. Specific gravity of the sulfuric mix should be equal to or less than 1.84. This is important because varying the strength of the acids, temperature, time of reaction, or acid-cellulose ratio will create widely different products. Caution must be exercised here.

2. Carefully rinse the cotton in pure water until it has a neutral reading with litmus paper or pHydrion paper.

If the cotton is the least bit acidic, rinse it again.

In the past, protocols varied with different powder makers, but a typical way was to boil the cotton for days with several changes of water, then chop it finely and boil it again with sodium carbonate and several more changes of water. Or it might have been left for days in a running stream of clean water. The reason for this was to remove the slightest trace of acid since any remaining would make the guncotton unstable (and cotton fibers are hollow tubes that tend to trap the acid). Also, most earlier makers had a residency time of at least 30 minutes and had protocols/specifications for preparing the cotton, etc.

3. Dry the cotton (which is now nitrocellulose). Take extreme care to keep the temperature of the nitrocellulose below 212°F. Once dry, the nitrocellulose must be stored at a cool temperature and kept away from sunlight.

Low-grade (low nitrogen content) nitrocellulose, not suitable for guncotton, readily dissolves in a mix of alcohol and ether. High-grade (high nitrogen content) nitrocellulose dissolves best in acetone. If what you now have dissolves readily and completely in the alcohol-ether mix, then it isn't suitable for this process.

At this point, you may process the nitrocellulose for powder by using acetone to dissolve it into a gelatinous blob. The addition of barium and potassium nitrate will increase the material that is burned, reducing fouling in the process. The slurry should be rolled into sheets and cut to the proper size and then redried. (It is also possible to mold the material through a pasta extrusion device to create fibers that can be dried into powder.) As with other powders, smaller particles are used for pistol ammunition and larger for rifles or shotguns. Working on Teflon surfaces simplifies matters.

To control burning rate and reduce barrel erosion, modern powders have deterrent coatings applied to them. One common coating for nitrocellulose is dibutylphthalate. Graphite coatings also are generally found as a glazing on ball powders; this glazing makes up 1 percent of the weight of powder and appears to have a slight deterrent effect. The graphite also reduces the amount of static electricity generated when grains of powder flow through powder measures and the like. Diphenylamine used to be added to powders as a stabilizer and is ideal for use in homemade powders.

Dinitrotoluene (DNT) is also used as a coating on powders. This chemical can reduce the initial burning rate and thereby reduce chamber pressures, making it ideal for many cartridges. The addition of potassium sulfate will often help reduce muzzle flash.

MODERN DOUBLE-BASED POWDERS

Double-based powder is even more dangerous to make than nitrocellulose. Consequently, you shouldn't even try the following manufacturing techniques unless you are a trained lab technician and equipped with modern lab equipment.

Modern double-based powder is made from nitroglycerine; do not attempt to manufacture smokeless powder unless you are in a position where the risks involved in not having ammunition makes the danger acceptable. Even then, extreme care must be taken; the desired results will not "happen" if you don't have access to a well-equipped workspace.

This powder is created from nitrocellulose and nitroglycerine. For making nitroglycerine, you'll need glycerine as well as the nitrocellulose, sulfuric acid, and nitric acid.

If you are equipped to create a modern smokeless powder, your first step is to create nitrocellulose as outlined above. That done, you'll continue as explained below.

But remember that these procedures are extremely dangerous because sulfuric and nitric acids are used. Both are highly caustic and can be contained only in glass without creating dangerous reactions. They are hard to neutralize; a spill will call for a lot of water and a strong base to clean things up. Gloves, safety goggles, lab apron, and hooded ventilation systems are required for this work.

Warning: The next steps will result in the creation of nitroglycerine, which is an extremely explosive material. Make only small amounts and do not shake the mixture or allow its temperature to rise above 86°F or else an extremely large explosion will likely result. Further, the mixture creates its own heat, so simply keeping it in an environment with below-86°F temperatures does not guarantee safety. *Note on materials for making nitro*: All materials used in bench-making nitro must be reagent grade and free of water. Traces of acid or impurities in the final product make it unstable and sensitive. Lab hygiene, care, and precision are very important.

After making nitrocellulose (as just described), the procedure for making smokeless powder is as follows:

1. Mix 100 parts of nitric acid with 200 parts of sulfuric acid in a water bath so that the containers will remain below 50°F. These acids create heat when agitated so take extreme care to maintain the mixture's temperature. Note: Both the nitric and sulfuric must be close to 100 percent. To concentrate the nitric, slowly boil water off until you get brown fumes of nitrogen dioxide. To concentrate the sulfuric, boil it until you get clouds of dense white fumes and the acid is a deep reddish brown. After the acids are mixed as described, let them cool.
2. When the temperature of the acids has stabilized, *very* slowly add 38 parts of glycerine. If at all possible, allow the glycerine to trickle down the glass tube rather than dropping it down into the mixture (which could cause a violent, explosive reaction).

   A flat bath works fine, and a good technique for adding the glycerine is to use a spray atomizer. Stirring the acids does not create heat—the creation of nitroglycerine is itself exothermic. Consequently, it is very important to closely monitor the temperature of the acid bath as you *slowly* add the glycerine; the faster you add glycerine, the faster the bath will heat. However, there is a point of no return, where, if you add glycerine too rapidly, the mixture will heat too fast and will explode.

3. Very slowly stir the mixture with a glass rod for 10 seconds. Stirring can create heat, so make sure the temperature is kept below 50°F, or else a violent explosion may result.
4. Gently pour 340 parts of water into the mixture *very, very slowly*.
5. The nitroglycerine will now precipitate out of the mixture. Carefully remove the nitroglycerine from the solution, taking extreme care to keep it below 50°F and to *not* shake or drop it.
6. Very gently wash the nitroglycerine several times with water and use sodium bicarbonate or lime to neutralize any acid in the nitroglycerine. Failure to neutralize or remove the acid will cause the powder to have a very short shelf life. Keep the temperature of the nitroglycerine very low.
7. Nitrocellulose and nitroglycerine are now dissolved in a solvent and extruded or molded into the correct size to burn properly and create the correct pressure curve for the firearm used. One good mix is 58 parts nitroglycerine, 37 parts nitrocellulose, and 20 parts acetone.
8. While wearing rubber gloves, gently knead the chemicals until they are well mixed. At this point, 5 parts petroleum jelly may be added—to reduce barrel erosion. If available, carbazole (also known as diphenylene amide) or diphenylamine can also be added to the powder mixture to help stabilize it.

9. Extrude the mix through a sieve, screen, or pasta press so that it forms threads. The smaller the strands, the faster the powder will burn. Large-size threads are suitable for rifle ammunition, and smaller threads for pistols and shotguns. Extruding thin threads of this is pretty hard to do without making some equipment, though some people have had good results using a garlic press. Early cordite was made in sheets and then cut into squares: the thicker the sheet, the slower the burn; the larger the squares, the slower the burn. Sheets were about as thick as a 3x5-inch card and cut into squares typically 1/16 inch. To slow the powder's burning rate further, it was rolled in graphite.
10. Allow the threads to dry at a very low heat and then break then into very short lengths.

To control burning rate and reduce barrel erosion, modern powders have deterrent coatings added to them. One common deterrent used with smokeless powder is Centralite I (diethyldiphenylurea). Graphite coatings also are commonly found; these reduce the amount of static electricity generated when grains of powder flow through various metering machinery and therefore aren't essential to powder manufacture. Best of all, they slow the burning rate for added barrel life.

Once made, smokeless powder is relatively safe to handle but will be sensitive to heat and light. Powder should be kept in sealed metal containers and stored at cool temperatures.

3

Top Brass

Unless you're using muzzle-loading firearms (which is an option you might consider), you'll need brass cartridges to reload ammunition. If you're firing a bolt-action gun or a revolver, this isn't much of a problem because it's easy to keep the empties. And, by exercising care, this brass can be reloaded up to 10 or more times, provided you avoid maximum loads, trim the cartridge to length, and thin down the neck of the case if required.

With selective-fire and semiautomatic weapons, a good brass catcher makes sense, though it can increase jamming and therefore isn't always suitable for combat (but, arguably, its suitabilty increases as your reloadable brass decreases). The best brass catchers are those offered by E&L Manufacturing. This company makes brass catchers for the Mini-14, AR-15/M16, HK 91/93/94, Uzi, MAC-10/11/11-9, .30-caliber carbine, AK-47, and other popular firearms that the government doesn't want you to own. The E&L catchers cost about $25 each and will quickly pay for themselves if ammunition becomes scarce.

If you're in an area where soldiers, rebels, or criminals are active, you might also be able to scrounge ammunition left behind after gun battles. However, you need to be cautious because these areas may be under surveillance as one side tries to ambush the other. And you might also be removing evidence if the police are going to investigate such a shoot-out. So be wary before scampering in to collect brass.

When scrounging empties, you need to remember that others may be reloading rounds with such materials as with mercury fulminate, making the rounds dangerous to reload. You should also avoid aluminum-cased empties for the same reason. Steel-cased pistol ammo can be reloaded, but it's hard on your reloading equipment. Don't try to reload it.

If you're forced to reload cartridges with black powder, brown powder, or any of the hygroscopic powders mentioned above, then you should take pains to wash the cartridges as soon as possible. This will not only clear out the fouling in them, but will also prevent corrosion that will eventually ruin the brass if it is left dirty. Hot, soapy water and a small brush are the best bets to get the residue off cartridges.

Some reloaders report good results after placing cartridges in a tray and running them through a dishwasher. The only caution here is that the high heat many dishwashers use during their drying cycle could cause the brass head to lose its proper tempering. Therefore, caution should be exercised with any approach that exposes the cartridge to extreme heat.

Even after being carefully cleaned, brass used with black powder will turn very dark because the sulfur in the powder discolors the metal. This black coloration is merely cosmetic, and the cartridges should remain strong if they don't become corroded and haven't been used with mercuric primers.

A common trick on the battlefield is to booby-trap gear, and even bodies, so an enemy will be injured when inspecting them. This may become a common tactic if law and order really break down in our country. In such a situation, a case of ammunition or other goodies where they don't seem to belong should be left alone or you're likely to blow off an arm—if you're lucky. About the only way to ensure that loaded rounds you find in the field are safe is to actually strip them from a recently captured or killed criminal, soldier, or enemy. If you don't see the guy up and running before he's aced, don't be tempted to remove ammunition or weapons from his body, because they might be booby trapped.

Cartridges should be carefully cleaned to keep grit out of your reloading dies. Failure to do this will create small scratches on the die and greatly shorten its life, as well as increase the likelihood of stuck cases. The best way to clean cartridges is with a vibrating cleaner. But if you don't have one of these, you can often improvise a "tumbler" with a small motor. I've seen improvised tumblers that use a waterwheel powered by a stream, as well as a container of brass connected to a windmill.

Whatever method of tumbling brass you employ, the shells should be placed in some sort of medium that will absorb the dirt and grit. Traditionally, walnut shells and hulls have been used for this; some reloaders also report good results with corncob grindings. One more thing: don't tumble live shells; even if they don't set each other off, the powder inside them will grind itself into a smaller size, increasing the rate of burning and producing dangerous chamber pressures.

It's also possible to chuck a cartridge into a drill press or lathe and clean it with a piece of steel wool. This gives the brass a like-new finish but is very time consuming (you might want to consider relegating this job to an assistant if you have one).

You can also clean cartridges with some common cleansers. Diaper Wipes, sold in grocery stores or drugstores, work well for this task, as does rubbing alcohol or even soapy water. Just be sure the brass is dry before attempting to reload it. Avoid chemicals formulated for cleaning brass; these concoctions contain an agent that makes brass brittle—a potential disaster when it happens to a cartridge case.

Reload brass with lighter loads rather than maximum amounts of powder to prolong the life of the cartridge. In addition to their being kinder to the bolt of the firearm, minimum loads will give you a lot more reloads with each brass cartridge since it will stretch less. If you have a semiauto or automatic firearm, creating minimum loads can be tricky; but even with these guns you can

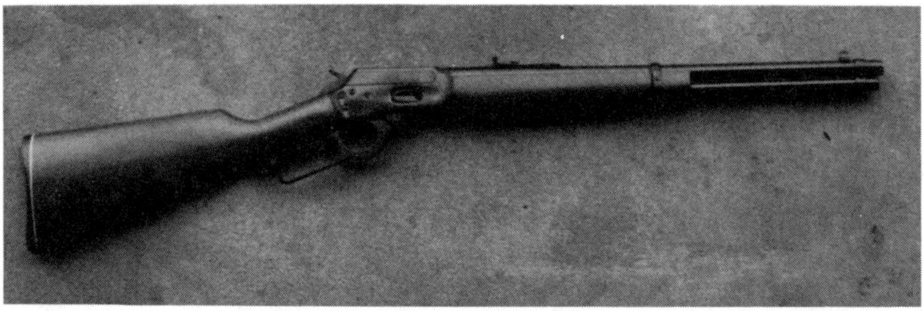

Manually operated guns like this lever-action Marlin are best suited to reloads that may be slightly underpowered.

Although most semiautos (like this Valmet M78 in .308 Winchester) require more precise pressure levels to cycle, it is often possible to reload at the lower end with mild loads that increase the life of brass and firearm alike. Careful cleaning of the barrel, gas piston, and bolt is necessary when firing such a gun with corrosive ammunition.

often reduce the powder charges somewhat and still have good dependability.

When brass has been fired and reshaped in the dies a number of times, it becomes somewhat brittle and may crack or split when fired. To get the brass flexible again, it's possible to anneal the brass neck to extend the life of the case. The catch is, that annealing must be done carefully, and the head of the case *cannot* be softened or it will become dangerous.

To anneal just the neck, you should first clean the cases (so you can gauge the temperature by color change). Once this has been accomplished, you should darken the light in the work area and place the cartridges in a pan of water with the water up to each case's shoulder. Each cartridge neck is then heated with a blow torch (or similar device) until the brass starts to darken.

The case should *not* turn cherry red; continue to heat the neck just three seconds or so after the case becomes dark. When this time has elapsed, knock the cartridge over so that the mouth is cooled by the water; this gives it the proper hardness for reloading. Take your time heating each case and try to spend the same amount of time on each one so that they'll all have similar softness. The beauty of this system is that the water keeps the case's head from becoming too soft. Again, you should be sure the cases are dry before reloading any ammunition.

If you have a large number of cases to anneal and have a lead-melting pot with a control to regulate the temperature of the metal, there's a better way to anneal the brass. Turn the temperature control on the pot to give a reading between 800° and 850°F and allow the lead in the pot to melt. Give each brass case a light coat of oil and then—while holding the head in your bare fingers—dip the neck of each cartridge into the lead up to the shoulder area. When the brass starts to feel warm, toss the cartridge into a container filled with water. (Avoid high-tin lead alloys such as linotype or Ludlow metals, wheel weights, or solder because they are more apt to solder to the brass cases.) Once your cases have been annealed and cooled, wash them with detergent to get the oil off the cases and then set them aside to dry.

Generally, you'll be able to anneal cartridges about two times before you start running into problems. These will give you a large number of reloads if you are careful to use light loads in them.

Don't try to dry cartridges in an oven; the temperature settings on ovens aren't overly accurate and permit the temperatures to rise high enough to ruin the cartridge's head strength—a situation that can occur at temperatures as low as 200°F.

After brass has been reloaded a maximum number of times, it can usually be safely fired as squib loads (practice loads firing very low-velocity projectiles). With care, brass can be reloaded an amazing number of times. The only time this isn't possible is with mercuric primers since they will cause brass to become brittle with time after being fired.

If you're really desperate for ammunition, it's possible to alter some ammunition to work in other chamberings. For example, the .30-06 or .308 Winchester cartridge can be cut down to .45 ACP length and the .223 Remington (and similar cartridges) can be cut down and expanded into 9mm Luger or .380 ACP cartridges. With a lot of work, you can even whittle a .223 cartridge down to .32 ACP or .30-caliber carbine size, though this gets to be quite a chore since the head diameter has to be modified through careful cutting in a lathe.

If a little careful bashing is done on a semiauto cartridge's rim, it is possible to peen the metal enough to create a rim. This makes it possible to cut and peen the .223 cartridge to create a make-do .38 Special or .357 Magnum cartridge. Likewise, the .308 Winchester can be transformed into a .45 Colt or, with careful work, a .44 Magnum cartridge.

It's important to remember with any modification of

rifle ammunition into pistol brass that it is a very crude adaptation at best. Further, the thicker brass of the cases requires that they *not* be loaded anywhere near the maximum capacity for standard cases: the reduced space inside the shell will likely result in excessive pressure and rupturing of the cartridge.

Consequently, you should employ such brass to low powder specifications, especially the first time it is used so that the firing will "fire-form" the brass to the chamber's dimensions. To make the rim of the altered case fit the extractor of the firearm, it may be necessary to place the brass in a lathe or drill and file it into shape (being careful not to weaken the case too much). An appropriate-sized E-clip snapped into the extractor groove of a rimless case might make a suitable rim.

Cutting the brass down to a shorter length is easily accomplished with a plumber's tubing cutter; a file and a brass trimmer will finish the work. When adapting the brass to a slightly larger size, it may also be necessary to make an expanding rod (as will be the case when going from .223 to 9mm Luger). Lubricate the cartridge carefully and send it into a resizing die on a press very slowly so as to give the brass time to expand without splitting.

None of the pistol brass created from larger cartridges will be overly reliable (especially in semiauto actions), nor will it be 100 percent safe, so you should try these procedures only in an emergency. Always use the proper brass whenever possible.

A few substitutions can also be made if cartridges in some chamberings are in short supply. The oldest standby is a .38 Special cartridge placed in the chambering of a .357 Magnum revolver—a practice that is regularly done and quite safe. With some guns, it is also possible to place a .357 Magnum into a .38 Special chamber; this isn't safe and will most likely warp or even rupture the chamber. However, if forced to choose between ruining the gun and possibly injuring yourself or dying at the hands of a hoodlum, you might opt to go out in a blaze of glory—and this just might be one way of doing that.

Of course if you're reloading, you can cut off a bit more than a tenth of an inch from the overall length of the .357 Magnum (so it will fit properly in the chamber of a .38 Special) and then reload it to .38 Special specifications. This will give you a round that will be both safe and effective in the .38 Special.

You could also reload .38 Special cartridges to .357 Magnum specifications, provided you don't go with maximum loads or try to fire them in a .38 Special revolver. This is potentially dangerous because the ammunition will fit in guns not capable of withstanding such pressures. So think long and hard before embarking on such a project.

It should be noted that the .38 Special +P loads fall between the .38 Special and .357 Magnum in power. Although most modern .38 revolvers will handle the .38 Special +P load, not all will, and many will shoot loose with a steady diet of these. So it is best to back off on charges if you're uncertain how tough the gun is that the round will be fired in.

Although it is fairly rare, the .38 Long Colt and (more rarely) the .38 Smith & Wesson cartridges will sometimes fit into .38 Special or .357 Magnum guns. And these older cartridges might escape the notice of those who would ban more common rounds since they are old and not very efficient ballistically.

For those with a .44 Magnum pistol, revolver, or carbine, the .44 Special can also be chambered in these guns. The round probably won't cycle a semiauto action, so be prepared to do this manually.

In addition to reloading cartridges, it is also possible to pull the bullet, remove the powder, retrim the case length, replace only some of the powder, and then reseat the bullet. This could be used to transform a .357 Magnum cartridge to a .38 Special or a .44 Magnum to a .44 Special. Just be careful to get the powder reduced properly or you'll be building a bomb rather than a cartridge.

Rimfire guns have the best interchangeability. The .22 Short, .22 Short CB, .22 Long, .22 Long CB, and .22 Long Rifle cartridges can all be fired in a gun chambered for the .22 Long Rifle. About the only problem these substitutions will create is a lack of cycling in semiauto guns. Extended shooting of shorter rounds in a .22 Long Rifle chamber might also create chamber erosion, but it would take quite a bit of shooting to do this, probably more than you'll see if you're forced to substitute ammunition.

None of these cartridges can be safely fired in a .22 Magnum gun. If you're willing to go to a lot of work, it is possible to wrap these cartridges in tape or paper to "jam fit" them into a chamber of a gun, turning it into a one-shot firearm until you can send a cleaning rod down the barrel to poke out the empty—a last-ditch situation at best.

There are some really dangerous substitutions that can be made from one type of chambering to the next, but these will often result in a ruptured case and escaped gas. These should be tried *only* if you will get killed without firing the round in the weapon. Firing a round in a different chambering is very dangerous at best. Although this could save your life if you need to fire the rifle in a combat situation, you should weigh the risks carefully before shooting.

One such example is the .38 Short Colt cartridge that will also chamber in a .38 Special gun. But this isn't overly safe to fire since the bullet diameter will create excessive pressure. The .30 Mauser and .30 Tokarev (7.62mm Tokarev) cartridges will also chamber in some .38 revolvers; these smaller diameter bullets rattle down the barrel, and the cartridges often leak gas around the cylinders as well, making them potentially damaging to the gun and shooter.

Other such combinations are the .223 Remington cartridge loaded into a .222 Remington Magnum rifle and the .308 Winchester cartridge loaded in a .30-06 chamber (which will sometimes work *if* the .308 is tight in the chamber). The brass in either of these make-do situations will be ruined and likely jam in the chamber to boot. Most likely, it will send a stream of high-pressure gas into the action of the rifle—and your face. You might take the risk hoping for one good shot if you find your back against a wall and not firing means death.

Those who've mixed .380 Auto shells in with the normal 9mm Luger cartridges have probably discovered that the two sometimes can be interchanged—though not with very effective or safe results. The key factor is whether or not the 9mm pistol's extractor will lock onto the rim of the .380 cartridge. If the extractor claw does latch onto the rim during chambering, it will hold the cartridge in place for the firing pin to reach the primer, firing the round. The bullet will exit the gun with a good velocity, though it is likely there will be extraction problems or jamming in the process. (Because of the longer length of the 9mm cartridge, it can't be fired in a .380 ACP pistol.)

Those who use a 9mm Luger revolver (like those offered by Ruger in its SP101 models) might also make the .380 Auto substitution for the 9mm Luger—again with some risk to both shooter and gun. In this case, all five rounds will probably fire, though the chances of extraction problems are great. And, obviously, a steady diet of such cartridges in the gun is likely to create chamber erosion. Again, this is only a last-ditch resort.

With shotguns, things are more limited when it comes to substitutions. Either .30-06 or .308 brass can be cut to remove the shoulders of the cartridge and a rim peened on the head to create a .410 shotgun shell of sorts. The .45 Colt cartridge can also be fired from many .410 shotguns (usually quite safely, depending on the condition of the shotgun). The .45 Autorim might work in the .410 in a pinch, and peening a .45 Auto shell so it has a make-do rim might also give you something that could be fired from the .410 shotgun. Of course, accuracy suffers with bullets fired from the smooth bore of the shotgun because the bullet tends to tumble rather than spin (perhaps a plus in terms of wounding potential).

Many .410 shotguns will also accommodate the .41 Magnum, .44 Magnum, or .44 Special cartridges as well as similar rounds. In the case of the magnum loads, it's wise to pull the bullet and reduce the powder charge to avoid damage to the gun and shooter. Of course, one could argue that really smart people would avoid these substitutions altogether unless there is no other choice.

The 12-gauge shotgun, like other rimmed cartridges, headspaces on the rim. This makes it possible to fire shells of varying lengths in these guns with some modern shotguns capable of handling everything up to 3 1/2 inches in length. This same feature could also be used to make more compact shells—if you want to go to all this work. Reducing the overall length of the shell from its standard 2 3/4-inch length to 2 inches, for example, would give you a few more bangs before reloading the magazine; this might be a consideration if you're using one of these guns for self-defense. Depending on the reloading mechanism on the particular gun, you should anticipate failures to feed if you shorten the case too much: loading mechanisms are designed to feed in shells of a certain dimension.

If you have access to a drill press or a lathe, you can create metal subcaliber devices from brass or aluminum that will accommodate a pistol cartridge, centering it in an adapter that has the outer dimensions of a shotgun shell. Of course, the pistol bullet will have a pretty low velocity when it leaves the shotgun since the barrel won't contain the gases behind the bullet. For this reason the "barrel" of the insert should be as long as is practical and exactly the same diameter as the bullet.

A 16-gauge shell can also be made to fit in the 12 gauge by wrapping the smaller 16 gauge with tape or paper; this is not safe, however, and 16-gauge shells are usually less common than 12 gauge, making this a doubtful proposition at best. Wrapping a 20-gauge shell and jamming it into a 12-gauge barrel's chamber might work, but it is also dicey.

I've heard stories about people who claim to have fired .50-caliber BMG cartridges from 12-gauge shotguns; I have to say I have serious doubts about such a practice. Though this would likely produce spectacular muzzle blasts, the velocity would have to be rather low, with more bark than bite, as the bullet and top of the brass cartridge rattle down the bore. Probably a better bet would be to pull the bullet from a .50-caliber BMG cartridge, cut the cartridge shoulder off (making it a straight-walled case), and then peening a rim on the head so it would headspace in a 12-gauge chamber (and maybe even catch the extractor claw on the bolt after the shell has been fired, if you're lucky). *A serious word of caution*: Don't try anything with the .50-caliber BMG cartridge in a shotgun unless you're facing certain death. Even then, you should realize that using this combination will most likely destroy your firearm and injure or kill you.

Loading the cartridge with a 0.729-inch-diameter bullet would give you a potent round because this is the inner diameter of a 12-gauge barrel. Obviously, you'd want a very soft-lead bullet if you had any choke on the shotgun barrel—the less the better with this or other slug loads.

A more practical way of firing odd rounds of ammunition in a gun not originally designed for them is to use cartridge adapters, which have been marketed over the years by Marble and other companies.

Harry Owen was one of the more notable successes in creating viable forms of these adapters. His operation has since been taken over by Bill Herrick, who sells the adapters through MCA Sports. Harry Owens' original units consisted of brass adapters that took up the slack in space between the chamber and a small cartridge; the newer units offered by MCA Sports are tougher and made of stainless or blued steel, but work basically the same way. The MCA Sports adapters could be worth their weight in gold—or ammunition—if the stores close for good.

To use an adapter, it first has to be loaded with the smaller cartridge that is to be fired from it. Then the adapter is chambered like a regular cartridge (easing the bolt forward by hand because releasing the bolt of a gun with a free-floating firing pin can sometimes fire a pistol cartridge in an adapter).

Once the cartridge in the adapter has been fired, the bolt is manually cycled and the adapter caught before the ejector can toss it into the grass. Reloading is a slow process; the empty has to be pushed out with a stick or other small object and a new cartridge placed in it. (With the .22 rimfire adapter, the procedure is much the same, but a small firing pin assembly goes behind the cartridge to transfer the force of the gun's firing pin from the center of the adapter to the edge of the rimfire cartridge.)

In addition to the lower recoil the smaller rounds offer, the report is also softer, making them ideal for quiet hunting or inconspicuous self-defense. MCA's blued steel adapters cost about $19, whereas stainless adapters run around $27.

The more common adapter chamberings permit firing .22 LR as well as .22 Shorts, .22 Longs, or .22 CB Longs in the .223; a similar adapter allows firing .22 Magnum rimfire cartridges in the .223. These adapters are also available for the .222 Remington chambering. For those having .30-caliber rifles, adapters that permit firing the .30-caliber carbine or the .32 ACP in the .308 Winchester, .300 Winchester Magnum, and .30-06 are also available. Other adapters are also available from MCA from time to time.

As far as actually making brass cartridges from scratch is concerned, the task is next to impossible for anyone other than a skilled machinist. The casting and swaging is a complicated process even with the proper dies and hydraulic presses.

You might create low-velocity, straight-walled cases with a drill press or lathe, however. These could be fired in a revolver, lever-action, pump, or bolt-action gun, but would be unsuitable for a semiauto or automatic weapon unless you were willing to chuck the rounds through it manually. You would have to keep powder charges in such rounds very low to keep from creating dangerous pressure if the powder capacity was appreciably diminished.

4 Best Bullets and Other Projectiles

Bullets are probably the easiest components to create—at least compared to the others already discussed. With a few good bullet molds or some swaging equipment, a person could make bullets with only a little scrounging around for materials. It's also possible to create bullets by turning brass or copper rods, though making these gets a bit tricky because they are armor piercing and therefore illegal to own in most areas. And even police officers who might turn a blind eye toward bootleg ammunition aren't going to do so if the rounds can penetrate their bulletproof vests. So you'd do well to shy away from these rounds unless everything has gone down the tubes in a big way. (For those interested in armor-piercing bullets, exploding bullets, and other exotic rounds, see my *Combat Ammunition of the 21st Century*, available from Paladin Press.)

Of course, if you're in a real bind, almost anything that fits can be fired down the barrel of a gun. Hardware stores are a great source for such ammunition: brass bolts, steel bolts, or metal screws can all have their heads chopped off and threads flattened with a file to be seated as bullets in brass cartridges. If a metal rod is a tad too small to fit snugly in the brass (which should be resized—by the firearm's chamber if nothing else is available), tape or paper can be wrapped around the projectile before seating it. Barrel wear is horrendous with such projectiles (I hesitate to call them bullets), but the risk might be worth it if you were completely out of bullets.

Plastic and wood have both also been used for short-range bullets. Plastic can be augmented with a lead insert to lengthen range; in fact, one of the most effective 9mm Luger rounds currently on the market is Federal's Nyclad, which is a soft lead bullet surrounded with a nylon jacket.

Plastics such as nylon are easy to work with, and a person could easily fabricate a mold for such bullets (or even use a mold designed for casting lead bullets). The big advantage of nylon as a jacket metal is that it is easy to work with but still strong enough to minimize leading in high-velocity pistol bullets (but don't try it in rifle bullets—it will stay behind in the barrel, creating a bore obstruction that will likely destroy the gun with another shot).

The possibility of a plastic bullet's shedding part of its jacket in the bore is always something to be kept in mind. Experiment by firing a number of bullets down the barrel, carefully inspecting for obstructions between firings to be sure the jackets haven't loaded in the bore.

The greater the plastic content, the faster the muzzle velocity of a plastic-based bullet and the faster the drop-off in its useful range. This can be useful in an urban area when you're trying to defend yourself but wish to avoid nailing some old geezer down the way. However, any projectile is dangerous, so don't think of plastic bullets as harmless. At close range, they are very dangerous.

Lead bullets are even easier to make than plastic ones and have the additional benefit of greater range. If you've had the foresight to get hollow-point molds, lead bullets are also very effective in combat. Wheel weights are a good source of lead for bullets. Although they don't expand as well as purer lead, the metal is hard enough to minimize leading, making the alloy a good trade-off with a minimum of worry about mixing in lead to make them softer. Space doesn't permit going into the details of casting bullets. But most melting pots and bullet molds come with detailed directions to get you started.

Of course, a bullet isn't just cast and crammed into brass cases. First it's sized (forced through a die), and, in most instances (depending on your machinery), it is lubricated at the same time. The grooves on the back of the bullet are for lube, which you can make yourself out of lamb fat and beeswax. Even if you don't have a sizer, you should still lube the bullets by standing them up in a flat pan and pouring molten lube up the level of the last groove. Lube goes a long way to help stop leading—that's why .22 bullets are lubed. And most bullet molds cast a rebated base on the back of the bullet, and this is so you can put on a copper gas check, which further helps stop leading.

The lead bullet will start to create leading problems when the muzzle velocity gets high. Leading starts to show up with the pistol cartridges like the .367 Magnum, .41 Magnum, and .44 Magnum, and will be especially bad with the 9mm if you try to obtain high muzzle velocities. The best solution is to go with light powder charges coupled with heavier bullets. Fortunately, Lee Precision and other companies offering bullet molds sell at least a few that will create larger, heavier bullets.

With rifle bullets, things get tricky. Usually the best route is to load cartridges with a minimal charge and a heavy bullet. This works best with manually operated actions, though some semiauto rifles may function with such a combination. Check the normal jacketed-bullet weight of cartridges loaded in the chambering you want to make lead bullets for and then check a reloading manual that lists charges for lead bullets. This done, pick out the mold that will permit casting the heaviest bullets that can be reloaded in that chambering.

If you boost the power of a cartridge to make a semiauto or selective-fire gun operate, then be prepared to spend a lot of time cleaning leading out of the bore—and don't do a lot of firing without cleaning the barrel thoroughly or you may create dangerous pressures when the barrel gets leaded up. A steel bore brush, solvent, and lots of elbow grease will remove the lead.

Because round-nosed bullets are notoriously poor when it comes to combat, you should consider using a drill press to add a hollow point to lead bullets you cast. Although expansion won't increase a lot with such a hole, it will at least make the bullet wound a bit more serious—and every little bit counts when it comes to handguns.

Of course, the best performance in rifles will result from lead cores with copper jackets. These can be produced if

Swaging dies make it possible for the reloader to create jacketed bullets with his reloading press. Shown here are a pair of .38 bullet-making dies from C-H (left) and a pair from Corbin for making .224-caliber bullets (right).

you buy a swaging die kit, available from Corbin, C-H, and other companies. Swaging dies are a bit tedious to make and carry price tags that start at around $40 and climb to several hundred dollars. Many also require a single-die reloading press to handle the high pressures needed to reform the jacket to the proper size.

If you're reloading .223 cartridges, Corbin's .22 jacket-making die kit is a good buy because it permits transforming spent .22 LR cases into .223 bullet jackets—with the addition of a lead core. (For a more detailed look at this process, see my *Combat Ammunition of the 21st Century*.)

Since jacketed bullets work well, many amateur ammunition makers toy with the idea of adding a coat of another metal to lead-cast bullets. This idea looks better on paper than in practice, and a few such coats can be dangerous. Tin or other metal washes should never be used because they can weld chemically to the brass cartridge over time and create excessive pressures when fired. One possibility that might work would be to add a copper coat through electroplating; a few commercial companies have tried this with .22 LR bullets and have had moderate success.

Many bullets are interchangeable with cartridges other than those for which they were designed. Weight can often be removed, with a little judicious file, drill, or knife work, to create a lighter bullet for smaller cartridges using the same bullet diameter. Lightweight bullets in heavier loadings usually are problem free, except for lead bullets, which may create leading problems if propelled too fast.

The .357 Magnum, .38 Special, .380 ACP, .32 ACP, and 9mm Luger (or other 9mm cartridges) can all use the same bullet *if* the weight is reduced for the smaller cartridges. Though the .357/.38 bullet is .002 of an inch larger in diameter than that of the 9mm and .380 ACP, the barrel will swage the bullet down to a good fit with only a slight gain in chamber pressures. Provided "hot" loads are avoided, this is a fairly safe practice. As with other possibilities in this book, bullet substitution can be dangerous, so think carefully before deciding to exchange bullets between cartridges.

The .308 Winchester, .30-06, .30-30, .30-caliber carbine, and .300 Magnums can all fire the same-diameter bullets (.308 inch), with the principal difference being weight; this bullet diameter will also work in the 7.62x39mm cartridge (which is usually seen with a .311-inch bullet), though some guns firing the round will have bore width a bit different from that of U.S. cartridges.

The .224-diameter bullet works in a huge range of .22 centerfire cartridges such as the .223 Remington, .222 Remington, .222 Remington Magnum, or .22 Hornet (or .22 LR, etc., if you're reloading it). Except for the .22 LR bullet (which is quite close to the .224 diameter), all these rounds use a .224-diameter bullet; again, weight and the avoidance of maximum loads are the main considerations.

Whether you're using the .308- or .224-diameter bullet, you can vary the weight of the bullet to the round you're reloading by drilling out lead from the nose of the bullet with a drill press and a very small drill bit. This can reduce the heavier 68-grain bullets designed for the .223 Remington to the 40-grain weight of the .22 Hornet, for example. Or you might cut off and round the nose of a 180-grain bullet designed for the .308 Winchester to reduce it to a 110-grain pill for loading a .30-caliber carbine cartridge.

Shotgun projectiles are readily improvised with small nails, screws, bolts, or even sand. There are several important considerations in improvised shot, however. One is that the lighter the material, the less its effective range. That means sand is going to be good only within 10 yards, in contrast to the normal useful range of 30 yards for lead buckshot. Likewise, the shape of the projectile will enhance or decrease its range; steel ball bearings travel farther with greater velocity than steel screws, whose shape quickly slows them down.

Another important consideration is barrel wear. Sending steel screws or sand down a barrel is going to scar it quickly and will eventually cause extensive damage to the smooth bore. The solution is to use a shot cup, perhaps improvised from plastic tubing or small plastic bottles. By encasing the abrasive shot in such a shot cup, the barrel will be protected and the shot pattern a bit smaller to boot, putting more of the projectiles on target. Shot cups will also help you avoid excessive chamber pressures, because odd-shaped projectiles such as screws could conceivably bunch up and bind in a barrel as they travel down it.

If you can obtain lead wheel weights easily, they would be ideal for making shot. By melting the lead and dropping it from the highest point available into water, you get a variety of pellet sizes. If the drop is long enough, you can even create relatively round shot (though probably with a lot of odd-shaped pieces as well).

Several companies offering bullet molds also offer molds for casting shotgun slugs. A slug transforms a shotgun into a weapon capable of reaching clear through a car body to strike an opponent and makes defeating lesser barricades a cinch. The slug also increases the useful range of a shotgun to 100 yards, making it a consideration for those facing an opponent with a rifle or pistol.

Whether you're using "bare" shot without a cup, a shot cup, or a slug, you can fill up the empty space in the shotgun cartridge with a variety of improvised materials; felt or cardboard discs are the most practical and are also easily made. It should also be noted that many shotgun shells were loaded with black powder during the first of the 1900s. This could still be done and would be especially practical with manually operated guns; a shotgun with a chrome bore is ideal if you're using corrosive primers or powder.

Your rifle and pistol cartridges will have a bit more shelf life and weather resistance if you run a small coat of lacquer or shellac around the bullet-to-cartridge seam and also around the primer. This will seal the cartridge against moisture and prolong its life. In the past, much the same thing was done with asphaltic varnish. This required that the inside neck of the case be coated with the material before the bullet was seated into it. You can create this chemical by thinning asphalt with a nonpolymerizing petroleum-based solvent; naphtha and acetone are good choices.

In general, lacquer produces better results, unless it is too thick, in which case it may create spreading problems. Thinning it to water consistency will avoid this.

Seating bullets on centerfire ammunition is pretty straightforward if you have a reloading press. With rimfire bullets, things are a bit trickier. The first thing to keep in mind is that the rimfire primer is easily set off during this reloading operation. For this reason, you should take special precautions; an explosion will most likely send shards of brass every which way and might even blow up a reloading die if you've created one for the cartridge. It's important to wear safety glasses with all types of reloading operations, and this is doubly true with .22 rimfire shells.

It should be noted that the .22 rimfire bullet is held in place by a crimp around the upper rim of the brass. This can be done with a small tool such as a pocket knife. Or you might want to just peen a few "holds" with a small punch, hitting along the upper edge of the brass to drive it into the side of the bullet.

Surprisingly, rimfire ammunition is more sensitive to moisture than other types of ammunition, so a coating of shellac or lacquer around the mouth of the brass where the bullet is seated is essential for .22 ammunition that contains hygroscopic powders or primer chemicals.

Shotgun shells are also very sensitive to moisture, thanks to their large primers and wide mouth area. The primer should be sealed with lacquer, as is done with rifle and pistol ammunition. The mouth of the case can be sealed with candle wax or lacquer as well. The main consideration in sealing the mouth of the case is to remember that too good a seal will create excessive chamber pressures that might split the case. You want a seal that keeps out moisture, but which is thin enough that it will rupture easily when the shell is fired.

5 Clean Machines

Because most of the powders and primers listed in this manual are hygroscopic or oxidizers, their use dictates careful cleaning of a firearm shortly after firing. Failure to do so can cause extensive rusting in the barrel or—in the case of gas-operated semiauto weapons—in the gas system of the rifle, as well as in the bolt and trigger group in guns like the AR-15.

Except for this headache, however, corrosive ammunition works just as well as modern cartridges. And a properly cleaned and oiled firearm will last as long (or nearly so) firing corrosive ammunition as with noncorrosive ammo.

If you're unfamiliar with the following cleaning procedure, you'll likely think it sounds like a good way to ruin a gun. But as many black powder fans who decide they know better have discovered, failure to follow the "standard" procedure of cleaning a gun of corrosive powder and primer residues will leave you with a rusty firearm in no time.

The secret to removing corrosive powders is to run boiling, soapy water down the bore several times, as well as into the action of the firearm or other areas where fouling is likely. A good copper or brass brush run up and down the bore, along with some elbow grease, will loosen everything for a final washdown with hot water. This is followed by cleaning the bore, which most shooters are used to doing to remove the last of the fouling. Then the gun gets a light coating of oil to prevent any rust.

Notice the temperature in the water specification above: boiling. The hotter the water, the better because the firearm will tend to heat up and the water will quickly evaporate off its surface, minimizing the chance of rust forming while you're cleaning it. However, you can clean a firearm even if you don't have hot water. If you're in the field, find a stream, pond, or even a mud puddle to supply the water to rinse out the firearm. Soap or washing soda, if you have either of these available, will also help loosen the potassium chlorate. (Soldiers on the battlefield have made do with oil, grease, or even saliva to clean guns of corrosive residue, but water is best.)

Careful oiling of the bore, interior, and exterior of firearms is all the more essential after cleaning them with soap and water, so don't skimp on this task once you've removed the corrosive powder. Modern lubricants that combine lubrication and bore-cleaning fluids simplify this task. A good bore brush and strong cleaning rod are also effective. If you've been firing lead bullets, consider investing in some of the new cleaning products designed to remove lead from the bore; they save a lot of elbow grease.

6

Caching In

Ammunition can be cached as easily as reloading equipment and firearms, and caching all three now would be wise. Who knows whether such items will become contraband?

You should realize that even though the government might turn a blind eye to some black-market activity, it isn't likely to ignore trading in such banned materials as ammunition. But you need to weigh the pros and cons carefully before setting out to disobey any law, no matter how unjust it may seem to you. It's obviously a lot easier to get arrested when you're doing things that are illegal than when your nose is clean (even though being innocent isn't always enough in this day and age).

Caching can be used to separate you from your contraband while still giving you access to it when needed. If you place your cache on property that isn't owned by you and isn't close to your home, then you've also created an extra safety margin that may enable you to stay out of jail.

There are two secrets to caching that, if remembered, will practically guarantee your success in preserving your reloading equipment and ammunition. One is that moisture and, to a lesser extent, heat are the enemies of almost anything you cache. The other thing to remember is that the cache must be positioned so its site won't be discovered by casual searches.

A poorly constructed cache is as bad as no cache at all. In the past, partisan fighters in Europe and the Philippines buried guns and ammunition improperly. When the weapons were retrieved to fight invaders, rust and moisture had made them inoperable, and the fighters were forced to battle with weapons in poor condition or those retrieved from the enemy.

Fortunately, several chemicals are available in most gun stores that will give added rust protection to the reloading equipment you cache. Outers' Metal Seal displaces water somewhat, adheres to the surface of steel, and acts as a lubricant as well. It's readily available at most gun stores and comes in an aerosol can, which makes application quick and easy.

For maximum protection, cached tools can be encased in paraffin, Cosmoline, or other grease or wax combination to prevent rust formation. This sealing mixture can be improvised by melting candles and mixing them with a heavy oil, taking care not to set the mixture aflame. A heavy coat of motor grease or even Vaseline is a good alternative. For small metal items, you can buy Outers' Gun Grease, which is formulated for long-term storage of metal tools and guns and is also sold at most large gun stores. (Aluminum handles or other aluminum parts found on presses don't need to be covered with oil or grease, provided they are kept dry, but all steel parts do.)

Before storing any metal object, clean its surfaces and wear rubber or cloth gloves while cleaning and packing the tool in grease so sweat doesn't get onto the metal and promote rust. (This will also prevent your fingerprints from being on the object if it is discovered.)

If you're caching reloading dies, be sure to pack the inside of the dies with grease; any rust here will severely damage the dies and perhaps even make them useless. If you're also putting firearms into the cache, two important areas to pack with grease during storage are the bolt and the bore. Be sure to coat the inside of the gun with grease as well as its exterior and cover the internal parts.

Once the dies, guns, and whatever tools you're caching are packed in grease, place them in a heavy plastic bag. The thicker the plastic, the better; and the less air in it, the less chance there is that rust or tarnish will form. Ziploc bags are handy for die kits, but use several of them around each die to ensure the moisture seal around them. The very best caching container is Bianchi Blue, a bag treated inside with a blue chemical that absorbs moisture and prevents rust as well. This bag is big enough to accommodate a large handgun and will work with much of your reloading equipment—though not the entire press. These bags are available at most gun stores or can be ordered directly from Bianchi International.

If you're pressed for time, simply force out the air in the bag with your hand and seal the bag. If you have more time, you can remove the air from a bag more thoroughly. One way is to use a small hand pump like that used to remove air from food containers. These are often available at health food stores and work well if you take a towel or other object to seal the plastic around the hose. You can also use water pressure to get the air out of a plastic bag (a method apparently conceived by Ragnar Benson—at least he's the first guy I know of to propose it). To do this, a bathtub or other container of water is required. Keeping the opening of the plastic bag well above the water, lower the bag into the water; the pressure of the water forces the air out of the bag. Then very carefully seal the opening and pull the bag from the water. Let it dry thoroughly before caching. Failure to let it dry completely will result in high humidity and rust inside the cache chamber.

Be sure the bag you're caching is sealed tightly. Ziploc-style fasteners are often not sealed completely, which allows air and moisture to flow in and out of them. Take extra care to double-check the seal.

Once the bags of equipment are sealed, they should go into a rugged container that will protect them from accidental breakage and prevent tearing of the plastic bags. This container will also help keep moisture out and protect the materials in the cache from insects and vermin.

There are two types of containers used in caching: an airtight container and an porous one that allows air movement through it. Although it seems like the porous container would promote rust, in fact it is sometimes a better choice, especially in dry areas, because it prevents moisture from accumulating inside the container. If your cache will be in an arid area, a porous container is better.

The porous cache box can be constructed of any type of wood, but exterior plywood is best. Use galvanized screws or nails to fasten the box together (regular nails or screws will rust, and the box will fall apart eventually). After the box is finished, paint it inside and out with urethane or spar varnish.

The watertight container is essential in humid areas. If a watertight cache is created, then it's necessary to place a hygroscopic chemical packet inside the container, away from the firearm or other metal tools. Silica gel is the hygroscopic chemical of choice and is available at most drugstores.

Polyvinyl chloride (PVC) pipe works well as a caching container. The pipe is strong but light, easy to work, readily available in hardware and plumbing stores, and also commonly used in household plumbing, so those selling it to you shouldn't suspect that you're making a cache. (For absolute security, purchase the pipe and fittings with cash to avoid leaving a paper trail.)

To create an airtight cache with PVC, simply purchase a length of pipe and two end caps for each cache, along with a can of PVC sealant or Volclay. Cut the pipe to length with a hacksaw or wood saw, then carefully sand the edges until they are smooth. Place the sealant on one end and mount the end cap, setting it aside to harden, following the instructions on the sealant can.

Now your plastic bags of dies and other reloading equipment can be placed into the cache tube along with the silicon gel packets. Styrofoam packing "popcorn" or similar material should be used to help protect things from rubbing against each other or creating damage. Some cachers use freon, nitrogen, or carbon dioxide gas to displace the air in the container; this helps prevent rust but isn't necessary. Of the three, dry ice (solid carbon dioxide) is the easiest to work with. You simply place a large chunk in the containers and then let it melt with the open top pointed upward in an area with very little air movement. Since carbon dioxide gas is heavier than oxygen, it pushes the oxygen out of the tube as it settles down into it. Once the dry ice melts, seal the tube immediately.

Sealing the tube is accomplished by either painting a strip of solvent around the top and then setting a cap on it, or placing the cap on the tube and sealing it with Volclay. A Volclay seal is easier to open in the field and even allows recycling of the tube; the solvent gives a more watertight fit but requires that the tube be broken or sawed apart to get inside it. You'll have to decide whether ease of opening or quality of seal is more important.

The size of the cache container is dictated by what you're putting into it. If you want to store a shotgun or rifle along with your reloading press, then the container will be larger than one that contains only a small reloading kit. It's better to have several smaller containers rather than one large one whenever possible. This makes caching them easier and also simplifies retrieving items after they've been buried. If you put everything into one large container that needs several people to lower it into the ground, then the chances that it can be retrieved by just one of your members (if he's the only one to escape) is smaller. Modest cache containers will improve your chances of avoiding detection when you're retrieving gear from the cache. And, of course, the old saw about putting all your eggs into one basket applies here.

If you're storing hygroscopic powder or primers, it's also wise to place them in a separate container or well away from your reloading equipment or firearms inside the cache. That way, if the chemicals manage to attract moisture despite your best efforts to keep it away, the dies or guns won't get rusty.

The cache should be located in an area slightly higher than its surroundings. This will prevent moisture from pooling over the spot where the cache container is buried. The container should be buried with a thick layer of rocks or gravel under it, permitting water to pool below the container and gradually seep into the soil. Roofing paper or plastic should then be layered over the container to divert water from the cache.

Once the dirt has been filled and tapped over the cache, the plants that were growing over the location should be replaced carefully to camouflage the position. You might even consider transplanting a bush over the cache to make it look like the earth has been undisturbed for some time. If the plant continues to grow, it will also discourage casual digging by anyone who wanders through the area.

If the authorities suspect that a cache is in the area, they'll probably use metal detectors to search for it. Some of the best metal detectors are capable of spotting a gun or other large metal object down to six feet or more and can be set to ignore small pieces of metal trash. But most of the equipment used by government agents won't be this sophisticated because it is too expensive.

These less expensive metal detectors can be foiled with very cheap countermeasures. Sprinkling bits of aluminum foil, BBs, tacks, or scrap metal around the area will do the trick. Unless the searchers want to dig up a huge area, they will most likely give up before locating you firearm. If you go to a little extra work and scatter metal over areas where nothing is buried (perhaps roughly in a ring around the actual cache—though nothing too obvious, such as putting the cache in the center of the circle, or the authorities might guess your site), those who might search and dig in a few spots to see if someone is trying to hide something will have nothing to show for their efforts and will likely give up.

In some areas, the soil has a lot of iron in it already, which can usually be recognized by the reddish hue of the earth. This soil is especially good for caching metal objects because it throws off less sophisticated metal detectors. Other such areas as junkyards, the backstops of rifle ranges, and places that collect metal in the ground are ideal. People with scuba or snorkel gear could hide caches at the bottom of lakes or ponds, though this would be a lot of work and the cache container would have to be *perfectly* watertight.

Much thought should be given to the placement of the cache to avoid having it in a location that is easily seen or frequently traveled. If someone spots you burying or digging up your cache, you're apt to run into trouble.

You should avoid places that are prime locations for future development. Returning to your cache site to find that a bulldozer has dislodged your buried treasure or that concrete now covers it can be quite a shock. Also be wary of places that already have roads built through them or various types of utility lines buried in them.

When digging the hole for your cache, place a tarp or blanket next to the hole to catch the spoil (the soil dug up during the digging). The spoil should then be carefully replaced into the hole and tamped down so that the area over the hole doesn't collapse when the earth settles, indicating that something is buried there. If you are careful with your spade work, you can replace the sod so that it is impossible to see whether anything has been buried. Any spoil left over should be carried away in the tarp and then spread around the area so that there are no telltale clumps of dirt left anywhere.

Bury the cache container at least four feet deep, so that a farmer working his field won't plow it up, and keep it away from trees since roots can work their way into cracks and break a container apart or ensnare it so that you can't get it out without dynamite.

It's impossible to overemphasize the need to keep careful records of where you locate your cache. Don't simply place it 10 paces from the pine tree; that pine tree may be gone the next time you visit the retreat. Grass fires, logging activities (legal or illegal), and farm development can alter the landscape so much that it's nearly impossible to recognize a site.

Don't place the cache near a prominent landmark. Landmarks attract visitors, and some of these people are apt to notice any depressions or other signs of digging. Landmarks also attract treasure hunters with metal detectors, looking for antiques or old coins. Consequently, landmarks should be used to give bearings to the cache, but not for the "X marks the spot" location of it.

If you can use a compass (and this is a skill you should work on mastering), then you can take compass readings to help guide your path. Ideally, you'd take at least three readings from various landmarks, giving you some margin for error as well as for the possibility that one landmark might be altered in the future by construction work or some natural occurrence.

Encoding some of the measurements on your map avoids giving away your cache's location if your map falls in the wrong hands. If you encode the reference points, be sure everyone in the group knows the code so that other members can locate the cache if something happens to you.

Once the cache is in place, avoid the temptation to "return to the scene of the crime" to check it. Leave it alone and only visit the area once a year or so to ensure that no one has dug it up. You should dig up a cache only to replace fuel or other perishable items or during an emergency to use what's in it.

When reloading equipment is removed from a cache, you may discover minor rust spots despite your precautions. Modest rust spots can be removed by lightly rubbing steel wool soaked in oil over the corroded spot. The metal doesn't need to be refinished if it is kept lightly oiled.

Finally, if you live in an area where the temperature dips below freezing for long periods of time, it may be impossible to retrieve your reloading cache during the winter months. You might be able to chip your way to the cache if you had to, but it would be quite a chore and would most likely attract unwanted attention. Therefore, plan on reloading ammunition only during warm weather when you can dig into the earth, and get things back into the cache before cold weather sets in. Having to use dynamite to dig a cache hole isn't exactly "low profile."

7

Never Disarmed

Those concerned that the gun grabbers might ultimately prevail should remember that even in prisons the inmates are armed, often with improvised firearms and ammunition. The most important tool for rearming yourself is between your ears. Until someone shouts "off with his head," you should be able to keep this important tool operational.

That said, things will be tough if our government manages to disarm honest citizens. But the success of disarming righteous men throughout history has been short-lived because moral people realize that immoral laws must be broken. Laws that prevent self-defense are among the most heinous and the most deserving of being ignored by otherwise honest people.

Thomas Jefferson in his *Commonplace Book* copied the words of Italian philosopher Cesare Beccaria, and these words are as true today as they were more than 200 years ago:

> *False is the idea of utility that would take fire from men because it burns, and water because one may drown in it; that has no remedy for evils, except the destruction of liberty. The laws that forbid the carrying of arms are laws of such nature. They who disarm only those who are neither inclined nor determined to commit crimes . . . serve rather to encourage than to prevent homicides, for an unarmed man may be attacked with greater confidence than an armed man.*

Elsewhere Jefferson declared, "No freeman shall ever be debarred from the use of arms." I hope that this manual will help you avoid being disarmed by a corrupt government.

Manufacturers and Publishers

Bianchi International
100 Calle Cortez
Temecula, CA 92590

C-H Tool & Die Corporation
106 N. Harding St.
Owen, WI 54460

Corbin
P.O. Box 2659
White City, OR 97503

DBI Books
4092 Commercial Ave.
Northbrook, IL 60062

E&L Manufacturing, Inc.
39042 N. Schoolhouse Rd.
Cave Creek, AZ 85331

Lee Precision, Inc.
4275 Highway "U"
Hartford, WI 53027

MCA Sports
Box 8868
Palm Springs, CA 92263

Paladin Press
P.O. Box 1307
Boulder, CO 80306-1307